EXPLORING CREATION WITH BOTANY

2nd EDITION

Jeannie K. Fulbright

Then God said, "Let the earth produce vegetation: seed-bearing plants and fruit trees on the earth bearing fruit with seed in it according to their kinds." And it was so. The earth produced vegetation: seed-bearing plants according to their kinds and trees bearing fruit with seed in it according to their kinds. And God saw that it was good.

Genesis 1:11—12

Exploring Creation with Botany
2nd Edition

Published by
Apologia Educational Ministries, Inc.
1106 Meridian Street, Suite 340
Anderson, IN 46016
www.apologia.com

ISBN: 978-1-946506-39-9

Cover and Book Design: Doug Powell

All Biblical quotations are from
the Christian Standard Bible (CSB)
unless otherwise noted.

Printed by Asia Printing Co., Ltd, Seoul, South Korea
March 2020

10 9 8 7 6 5 4 3 2 1

Apologia's Young Explorer Series
INSTRUCTIONAL SUPPORT

Apologia's elementary science materials launch young minds on an educational journey to explore God's signature in all of creation. Our award-winning curriculum cultivates a love of learning, nurtures a spirit of exploration, and turns textbook lessons into real-life adventures.

TEXTBOOK

Apologia Textbooks are written directly to the student in a highly readable conversational tone. Periodically asking students to stop and retell what they have just heard or read, our elementary science courses engage students as active learners while growing their ability to communicate clearly and effectively. With plenty of hands-on activities, the Young Explorer Series allows young scientists to actively participate in the scientific method.

REGULAR AND JUNIOR NOTEBOOKING JOURNALS

Spiral-bound **Apologia Notebooking Journals** are designed to enrich your child's learning experience. These companions to the award-winning Young Explorer Series enable students to personalize and capture what they have learned in an artful keepsake. The Junior version of the notebook is created with primary-ruled lines to help students who are still developing basic writing skills.

AUDIO BOOK

Some students learn best when they can see and hear what they are studying. Having the full audio text of your course is great for listening while reading along in the book or riding in the car! **Apologia Audio Books** contain the complete text of the book read aloud to your student.

SCIENCE KIT

Apologia has developed an exclusive **Science Kit** for *Exploring Creation with Astronomy, 2nd edition*. This kit contains materials needed to complete the textbook activities, as well as the additional activities found only in the Apologia science kit, such as completing an Engineering Design Process to plan out a planetary mission and building a prototype robot to complete that mission.

TABLE OF CONTENTS

LESSON 9:

GARDENING 133

LESSON 10:

TREES 151

INTRODUCTION TO BOTANY

digging deeper

I am the vine; you are the branches. The one who remains in me and I in him produces much fruit, because you can do nothing without me.
John 15:5

Living Book

Thank you for choosing *Exploring Creation with Botany, 2nd edition* as your science book this year. This Charlotte Mason styled living book will help you cultivate your love of learning in your homeschool as you study God's incredible world of plants.

Activities and Experiments

The *Exploring Creation with Botany, 2nd edition* course is designed to provide fun, educational, and safe activities for you to conduct in your home, yard, and surrounding environments. We have tested these activities in multiple settings. Accidents, however, can occur anywhere and at any time. Therefore, we urge you to follow safe experimental processes at all times. Be sure adult supervision is provided for all activities. Follow standard safety procedures by ensuring use of eye protection and gloves if needed. Set aside equipment to be used only in experiments. Never experiment and cook or eat with the same items, especially if you are using inedible or potentially toxic materials. You should NEVER eat any item in an activity unless the lesson instructions are for an edible activity AND your parents have read the lesson, checked the materials, and given permission.

Course Website

To get the most out of this course, you should regularly visit the course website. It is designed to link you to multimedia and websites that relate to botany. Please always practice safe Internet use. While we monitor our book extras, we do link to outside sources, and we cannot guarantee each site on a day-to-day basis. To go to the course website, simply type the following address into your web browser:

apologia.com/bookextras

Type the following password into the box:

Godmadeflowers

Make sure you capitalize the first letter and make sure there are no spaces in the password. When you click "*enter*," you will be taken to the course website.

Notebooking Journals

The 14 lessons in this book are in-depth and contain quite a bit of scientific information and hands-on activities. Each lesson should be broken up into manageable time slots depending on a child's age and attention span; this will vary from family to family. A suggested lesson plan is located in your Notebooking Journal. Whether you are using the regular or junior Notebooking Journal, this journal will be your place to document your studies in botany, record your activities, and create a wonderful keepsake of all your studies.

When you read this science book you are studying important information discovered by great scientists throughout history. Keeping a student Notebooking Journal helps to document studies in your own words. Additionally, participating in science activities builds upon that knowledge as you conduct your own scientific studies. It is our prayer that at the end of the year you will have gained a deeper appreciation for and understanding of the beauty of creation.

Make your ways known to me, LORD;
teach me your paths.
Guide me in your truth and teach me,
for you are the God of my salvation;
I wait for you all day long.
Psalm 25:4–5

LESSON 1

WHY BOTANY MATTERS

digging deeper

Then God said, "Let the earth produce vegetation: seed-bearing plants and fruit trees on the earth bearing fruit with seed in it according to their kinds." And it was so. The earth produced vegetation: seed-bearing plants according to their kinds and trees bearing fruit with seed in it according to their kinds. And God saw that it was good. Evening came and then morning: the third day.
Genesis 1:11–13

All animals that live in the rainforest depend on plants for survival.

WELCOME

As you embark on this year's science course, discovering the secrets of all God created on the third day, remember that the Earth is the Lord's and everything belongs to Him—including you!

Your journey into the wonderful world of **botany** begins right here, right now. All that you learn this year will be important for the rest of your life. Why? Well, let's find out.

Botany, as you probably know, is the study of plants. It is one of the most important fields of science in the whole world. Why? It's because our very survival depends on plants. You see, plants produce the air we need to breathe, the food we must eat, the medicines that make us well, and many other items we use every single day. In fact, this book you are reading is made from plants! And let us not forget that plants also make Earth a more beautiful place to live. Imagine a world without any plants. What would you eat? Think about that. Did you know that virtually everything you eat is either a plant or something that depends on plants for survival?

Humans are dependent on the food crops that farmers grow such as this orange grove.

Think about pizza, for example. Every part of a pizza requires plants to exist. The crust is made from wheat. The tomatoes sauce is made from the

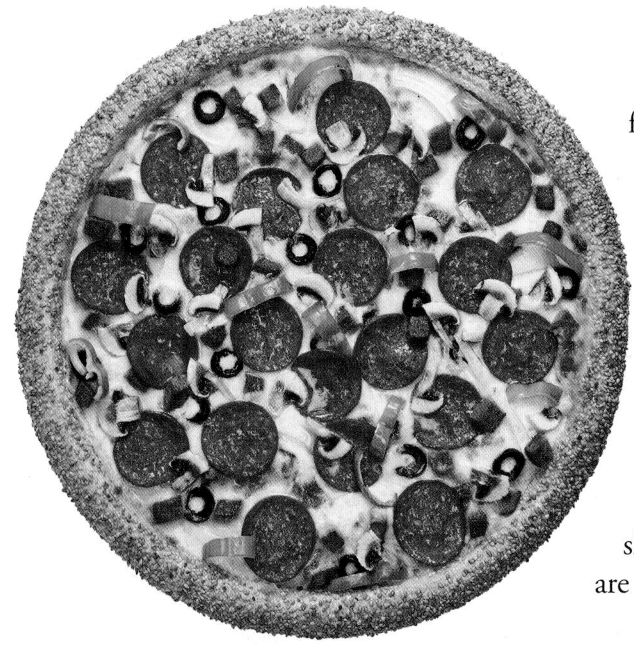

This pizza could not exist without plants.

fruit of the tomato plant. Several other fruits and vegetables add to the taste and flavor. But what about the pepperoni? It's made from pork and beef. Pork comes from a pig, and beef comes from a cow. Well, can you guess what pigs and cows eat? Plants! Without plants, there is no pork or beef. Indeed, almost every animal you eat—such as a chicken and fish—lives on plants. Of course, some people eat alligators, frogs, and such. However, even though those animals eat other animals, the animals they eat actually live on plants. So it all comes down to plants. We need them. And we should learn as much as we can about them because they are a vital part of life on Earth.

By understanding the secret world of botany, you'll become an expert on what it takes to grow and nourish plants and help them flourish in this world. When you're done with this first lesson, you will create a special structure called a light hut that's designed to grow plants from seeds any time of the year. You can plant any seeds you want, but I recommend herbs if you're starting in the fall because herbs can be grown indoors through the winter. If you are beginning in the spring, plan for fruits and vegetables because you'll be planting an edible garden outside! That means you'll grow food like strawberries, blueberries, carrots, pumpkins, tomatoes, lettuce, and so much more. If you live in the southern United States and are starting this book in the spring, you may want to go directly to Lesson 9 to begin building your outdoor garden.

Herbs are easy to grow indoors and can be used for cooking.

In addition to these important projects, you'll do experiments, dissections, activities, and much more. As you can see, you're going to learn a lot of important information this year, and I'm quite certain you'll be a brilliant botany student by the end of this course!

But botany isn't all we have planned for this year. In addition to learning about plants, you'll also learn about **fungi** (fun' jye). What on earth are fungi? Have you ever seen mushrooms growing outside? They are fungi. Because fungi also grow outdoors alongside plants, sometimes even helping plants, we're going to take one lesson to dive into the fascinating world of fungi. I think you'll find this special creation of God quite interesting.

This mushroom is a fungus.

CREATIVE CREATOR

Just as you do when you create things, God cares about His creation.

Genesis 1:11–13 tells us that God created plants on the third day of creation. And boy oh boy, did He do an amazing job! In fact, we might say that plants truly magnify the Lord. Are you wondering what it means to magnify the Lord? Have you ever looked through a magnifying glass? A magnifying glass makes things look bigger so we can see them better, and that's exactly what God's beautiful world does. It magnifies God so we can see Him better. When we study the flowers—their beauty and how perfectly they were created—we know that God is also beautiful, perfect, and creative. But of course our Creator is creative! He used unbelievable imagination when creating Earth. Did you know that God even gave you the gift of creativity when He designed you in His image? Do you have an imagination? Of course you do! You were made in the image of God. Did you know that it magnifies God when you do creative things? It certainly does. That's because you're using the gift of creativity He gave you. What are some creative things you like to do? Do you like to draw pictures, tell stories, or build things? Do you imagine new adventures or design new ways of doing things? Don't ever forget that God loves you and specially designed you with your own unique way of expressing your creativity.

In addition to creating the wonderful person that you are, God created so many different and amazing kinds of plants. In fact, there are thousands of unusual plants in God's kingdom. For example, some plants capture insects or small creatures and consume them for food. One kind of plant grows a flower that's three feet long (probably as long as you are) and smells like rotten meat. Some flowers love the sun so much that they turn around throughout the day to always face it. Other flowers don't like the sun, staying closed during the day and opening only at night. A few flowers are specially designed for only one single kind of animal to drink their nectar. God even created flowers that look like animals. Did you know that a special tree in Madagascar drinks in water, storing it in its trunk like a giant water bottle? God created some trees to grow so tall they seem to reach the sky. Would you be surprised to learn that the oldest living thing we know of on Earth is actually a plant?

You can see from this Madagascar tree's swollen trunk why it's named the water bottle tree.

The design of these flowers is perfect for giving the hummingbird food.

Birds will feed on the nectar in these bleeding heart flowers.

As you journey through this book, you'll learn about soil, seeds, roots, leaves, flowers, creatures that help plants grow, trees, and so much more. You'll do projects like create a miniature indoor greenhouse (the light hut I mentioned earlier)and make a field guide of plants in your area. You'll do experiments with plants, like changing the color of flowers. You'll also record what you learn in your Botany Notebooking Journal. Very soon, you'll make a special book called a nature journal and, like a true scientist, you will record your scientific observations of the outdoor world on its pages. I think you'll enjoy digging into the natural world of plants and fungi, don't you?

Now let's learn about nature journaling and get started on our scientific studies.

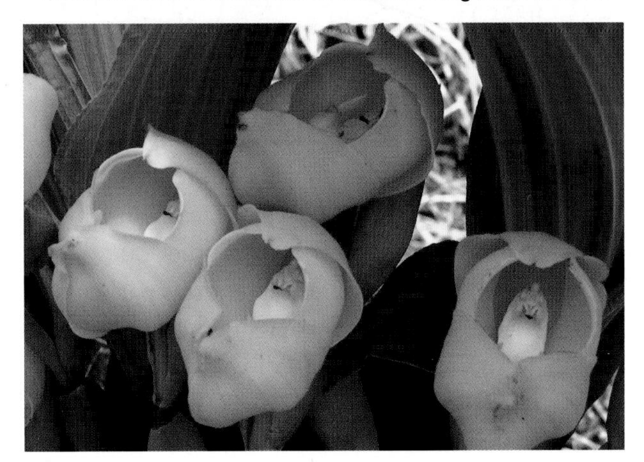

This cradle orchid is sometimes called a swaddled baby orchid.

NATURE JOURNALING

Throughout history scientists, naturalists, inventors, and explorers have kept detailed accounts of what they see, experience, wonder about, and learn. Most of these people put their thoughts, ideas, and observations in a special book called a log or journal. Much of what we know about the world today we've learned from these scientists and explorers who shared their books with us. People such as Leonardo da Vinci, Lewis and Clark, Alexander Graham Bell, and James John Audubon left us amazing journals filled with their drawings and writings. These recordings have taught us a great deal about nature, inventions, territories, cultures, and human beings. We see from their journals how they took the time to study with great care what they saw.

Leonardo da Vinci made many kinds of sketches in his notebooks, from technology and animals to human anatomy. This is a page from one of his notebooks.

Usually, when we go outside, we don't take time to focus on the details around us. But if we do this, we will learn the skills of observation that gave them such great insight.

Like many great scientists from the past, you will keep a journal this year. It will be a nature journal. To start, let's try an activity that will help you become more observant so you will be a better nature observer and nature journalist.

ACTIVITY 1.1
THINK LIKE A SCIENTIST

Go outside in nature and begin looking around. Notice a plant as you walk past it. Take a leaf from that plant and look at it very closely. Notice its shape. How does it feel? Is it soft or hard? What color is it? Do you know what color it will be in another season? Do you know if it survives through the winter or if it will fall from the plant and die? How big is it compared to other leaves? What kind of plant does it grow on? A tree, a bush, a vine? Think about the answers to these questions and come up with other questions about it. Make your own unique observations about this particular leaf. If you have a magnifying glass, see if there is anything more to observe that you didn't notice when looking with your naked eye. Save your leaf for the next activity.

During the last activity, were you surprised at how much detail there was in that single leaf? Do you now see the plant a bit differently, understanding a little more about it than you did before? Even if you don't, that activity helped you observe a leaf the way a scientist does. You were training your eye to see and your mind to think critically, with more attention to detail, just like a scientist. That's what we will be doing a lot of during this course. By keeping a nature journal, you'll spend some of your time outdoors thinking hard about what you see. Instead of frolicking around the whole time, playing and enjoying the outdoors like a child, you'll take some time to enjoy the outdoors like a scientist.

So what exactly are you going to do with your nature journal? You'll take it on walks and hikes or to botanical gardens and nature nurseries. This journal will become your very own record of the things you see in nature—the plants and animals you find interesting.

The next time you are outdoors, I want you to slow down and pay close attention to what is around you. You'll be training your eyes to see, teaching your ears to hear, and schooling your mind to be still so you can notice the smallest details around you. It's when you begin to really see the elements of nature that you start thinking like a true scientist! After looking carefully at nature, you will begin recording your observations.

ACTIVITY 1.2

MAKE A NATURE JOURNAL

You will need:

- Cover paper (construction paper, colored card stock, or scrapbook paper)
- Copy paper (10–12 sheets)
- Stapler
- Stack of cardboard (or substitute a kitchen cutting board)

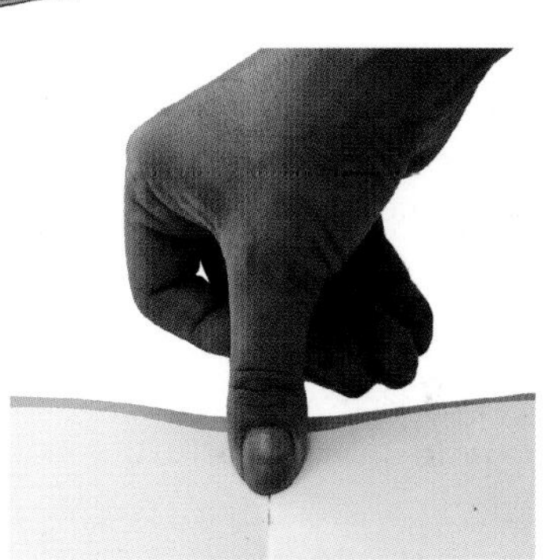

You will do:

1. Stack your copy papers on top of the cover paper.
2. Fold the papers in half and make a crease down the middle.
3. Turn the pages over and lay them flat on top of the cardboard stack with the cover paper facing upward.
4. Open the stapler and position it so that the staples will align with the crease in the center of the cover paper.
5. Staple three or four staples down the center of the papers to create the journal.
6. Pull the journal off the cardboard.
7. Turn it over so that the inside middle pages with the staples are facing upward. Ask an adult to flatten the staples in the center (You can use a butter knife to aid in this).
8. Decorate the cover however you wish. You may want to glue the leaf you studied earlier to the cover or make a drawing of it.
9. After you fill this journal with your thoughts and observations, just come back and make another one!
10. You will create a special pocket to store your nature journal inside the cover of your Botany Notebooking Journal.

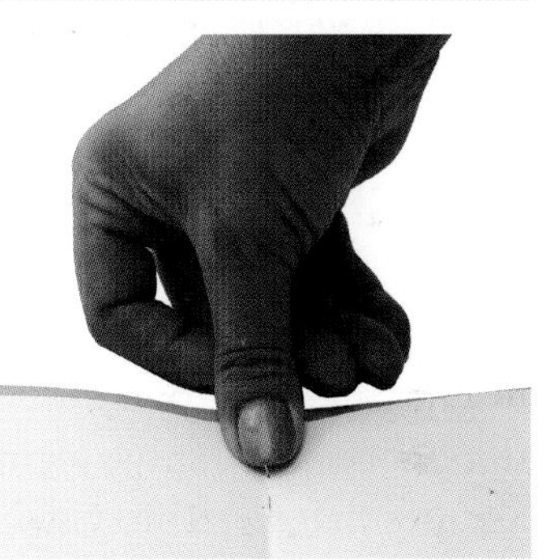

USING YOUR NATURE JOURNAL LIKE A SCIENTIST

ILLUSTRATIONS, SKETCHES, AND DIAGRAMS

Illustrations are drawings. You can illustrate a part of something you see, such as a petal. Or you can illustrate the whole plant, or even the entire scenery, such as a waterfall and all the rocks and trees around it. Don't forget to bring a set of colored pencils for your illustrations. It might be helpful to get a book on how to draw. This will help improve your illustration skills more quickly.

Sketches are a little different from illustrations. They are often drawn quickly with only a pencil. You can also diagram what you observe, including labels with lines and arrows pointing to different parts of your sketch.

DATE, TIME, PLACE, AND WEATHER

The most important information to put on each page of your journal is the day, the time, and the place where you are observing nature. Be sure to include the time of year, the temperature, and what the weather is like that day.

Remember, good science begins by asking questions. This will be your first scientific inquiry for this course. It is my hope that sometime during this course you will discover the answers to your questions and learn secret things about the plant world that you didn't know before.

SPECIMENS

Very occasionally, and with your parents' approval, you can include in your nature journal samples of the leaves and bits of nature you find outdoors (as long as they are not too bulky). However, be careful not to disturb too much of nature when taking a specimen. You wouldn't want to remove the only flower in a field. You could disrupt the natural cycle of that plant's growth.

Press flowers for several days between two heavy books to flatten them before adding them to your nature notebook.

DESCRIPTIONS AND THOUGHTS

In their nature journals, scientists make detailed notes about what they see. They observe things carefully and notice small changes and differences in nature. In order to journal like these scientists, you'll need to make yourself think very carefully, asking questions about what you are observing. Is there anything special about the plant or animal that you notice? What does it look like? Does it have any special features that seem interesting or different from other plants or animals? If it is a plant, can you describe the shape of the leaves, the look of the stem, or the number of petals on the flower? What color is it? Look at it through a magnifying glass. Describe what you see in your journal. You may also want to write down what you think about while you are outside in nature. What can you note about the day, the weather, or the time of year?

LISTS

Sometimes it is helpful to make lists of things you see in nature. Simply write down every kind of tree you see on your nature walk through a certain park or record the name of every bird that comes into your backyard on a particular day.

Your lists will prove to be an interesting and important aspect of scientific study. With lists, you can begin to notice things that change in your area. Perhaps you remember always seeing a certain flower or bird in the spring, but later on you notice that it isn't present anymore. These are the kinds of observations that genuine scientists make.

ACTIVITY 1.3

JOURNAL ABOUT NATURE

Go outdoors and spend some time closely studying the plants in your yard. Be sure to ask lots of questions as you consider what you are seeing. Even if you don't know the answer, questions are a great way to begin thinking like a scientist. When you find something that seems interesting to you, illustrate it and write about it in your nature journal. Be sure to record the date, time, and place where you observed it.

SCIENCE OF BOTANY

You are certainly off to a great start in your science studies this year. I hope you are getting the knack of closely observing nature and recording your findings in your nature journal. Let's talk a little more about the science of botany.

VOCABULARY OF BOTANY

As it is with all sciences, you will learn the special vocabulary for the field of botany. Most often, the words you will learn come from ancient Latin or Greek languages. Since those languages are no longer spoken, they are great choices for science words. That's because the meaning of the words will never change. You see, in languages that are still in use, meanings change quite a bit. For example, many years ago, the word awful meant something that inspired a sense of awe. However in the language of our day, the word awful means something that is terrible. And we now use the word awesome to mean a sense of awe. So that's why you will find a lot of Latin and Greek words in science. They don't change and you will never be confused about what they mean. Don't worry about remembering all the vocabulary words you learn. The most important thing is that you develop a strong understanding of the science of botany and build memories of your learning through lots of hands-on experiences. Nevertheless, I am going to teach you a lot of new words this year. Let's take a look at some of these special words and find out more about the science of botany.

This is a Latin inscription on an ancient stone.

Botanists are scientists who specialize in the study of plants.

BIOLOGY OF BOTANISTS

Did you know that botany is a biological science?

What's a biological science? Let's take a look. Have you ever heard the word **biology** (by ahl' uh jee)? It sounds like a complicated word. In Greek, **bio** means life and **ology** means "the study of." So biology just means "the study of living things or life." Plants are living things, so botany is also biology. A botanist is a biologist who studies plants. So if someone asks what you are studying in science this year, it would be correct to say you are studying biology. Zoology is the study of animals. It's also a biological science. So is human anatomy. There are many other fields of biology. After you learn botany, perhaps you'll want to study another biological science.

So what do botanists actually do? Well, have you ever taken medicine that healed you of an illness? Some botanists study plants that are used to make medicines that cure diseases. In fact, many different medicines are made from plants. Before modern medicine, people relied on healers who knew which plants helped to cure which diseases. They weren't officially called doctors, nor were they called pharmacists. But they were actually both. A doctor figures out what's wrong and then prescribes medicine that a pharmacist prepares for the patient. Many years ago, the healer figured out the problem and prepared the medicine using mostly plants. For example, if someone came down with a bad cold, a healer might make a tea from the sage plant. Sage is an herb that has many healing properties. Not only does it help with colds, but it also fights bacteria and helps with breathing problems. In fact, its scientific name is *Salvia Officinalis*, a word which means "the plant saves people."

For thousands of years, herbs have been used for healing as in this sage tea.

Hundreds of years ago, healers had a vast knowledge of botany. Even today, many people use plants to help with healing. When I was a child, my mother would slice open an Aloe vera cactus and rub the pulp on my sunburn to heal it. Today, I use oil from the tea tree plant to heal my children's cuts and keep the mosquitoes away by rubbing it on our skin. Chamomile is an herb that helps stomach pain, fevers, colds, and asthma. It also helps people sleep. Echinacea is used to fight off a cold. Many people use oils from plants to help with different complaints and problems. However, most of us do not rely on plants anymore to heal us directly.

Today we rely on pharmacy companies to make medicines, and many of those medicines come from plants. Why would we rely on others when we could use the plant ourselves? Because

The leaves of the Aloe vera plant contain healing properties and are especially helpful for burned skin.

modern medicine has taken the science of healing plants and improved it in many ways. Instead of using the entire sage plant, a pharmacist might determine what element of the plant is helpful and try to isolate that element to put into a pill form that people can take. So pharmacists use botany to make medicines and find cures for diseases.

Some botanists experiment with plants to learn more about them and how they can make specific crops grow faster, stronger, and better. They hope to produce plants that will grow more fruit that is bigger and more nutritious. They want the crops that we eat, like corn and wheat, to resist disease and be less attractive to pests. In areas where water is scarce, botanists attempt to produce crops that need less water. There are so many ways botanists can help farmers and improve the foods we eat.

As you can see, there are many ways botanists help the world. Since plants are so useful and essential to humans, the study of botany will always be a very important field of science.

Before we look at how a biologist organizes the world, tell someone everything you have just learned about botanists.

TAXONOMY

Do you like to have things organized? Do you like it when all your shirts are in one drawer, your socks are in another drawer, and your pants are in yet a different drawer? It makes life a lot easier when we are organized. Well, biologists like to organize things too. Living things. God created so many different living things that biologists have spent a great deal of time separating them into different groups, called kingdoms. This helps biologists keep things organized. They've put plants in one huge group, called the plant kingdom, or **Kingdom Plantae** (plan' tay). They've also put mushrooms and other similar growths in their own group called **Kindgom Fungi**. We will study these two kingdoms this year. But there are other living things, aren't there? Can you think of a living thing you've seen recently? You might have mentioned some kind of animal. Maybe next year you will begin studying the creatures in **Kingdom Animalia** (an uh

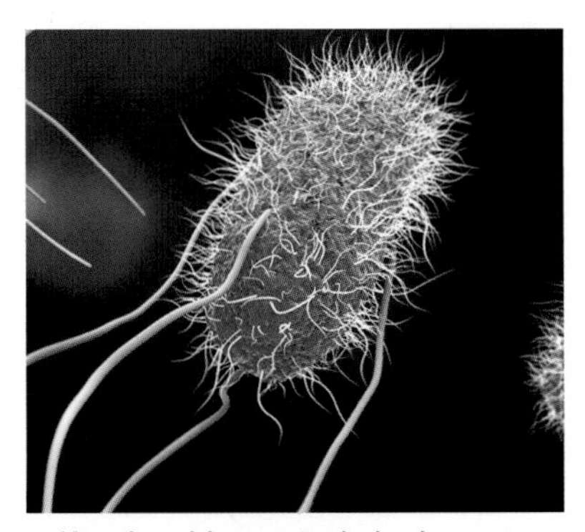

Kingdom Monera includes bacteria.

Kingdom Fungi includes mushrooms.

Kingdom Protista includes algae.

Kingdom Animalia includes all the animals.

mal' ee uh), otherwise known as the animal kingdom. But there are other kingdoms as well. Some kingdoms involve life that is so small that you can only see it with the help of a microscope!

But the organizing doesn't stop after dividing up the plants, animals, and other living things. Nope. Biologists like to place living things into even smaller groups. Let's take a look at a few of the ways they divide up plants.

VASCULAR PLANTS AND NONVASCULAR PLANTS

There are two main groups of plants, vascular and nonvascular. Before we discuss these two groups, I want to ask you a question. After you wash your hands in the sink, do you know where the water goes? Do you know where it comes from to get into the faucet? Well, it travels through tubes that are inside your house. Those tubes, or pipes, bring the water in and take the water out. You might say water is transported through these tubes. Guess what? That's one of the things botanists look at when dividing up plants: whether they have tubes in them or not. The plants with tubes are called **vascular plants** and the ones without tubes are called **nonvascular plants**. The word vascular just means a hollow container. The tubes are vessels inside the plant that transport fluids, like water.

Behind the walls of your house are pipes and tubes where water and gas flow.

You're probably trying to imagine tubes in a plant and what they look like. Do you think you might have tubes inside you? Look at your wrist. Do you see blue streaks? Those are tubes called **veins** (vaynes). They contain fluid. What do you think that fluid is called? It's called blood, of course! You see, then, that you and I are vascular, similar to vascular plants!

Inside your body are tubes filled with blood.

This leaf skeleton shows us the veins that are under the surface of the leaf.

As you'll discover over and over again when learning science, God made many living things in a similar way; plants and people both have tubes inside them. This shows us that God is consistent in His design of the world. Think about the pictures you draw. All of your pictures are similar to one another because they are made by the same artist—you! This is how it is with God's world. We can see that everything in the world was made by the same artist—God!

Look at the leaf in the picture. Do you see the veins? These veins carry water and other important nutrients throughout the plant. Although you can actually see many of the tubes in a plant by just looking at its leaves or flowers, many are also hidden inside the plant and are hard to see. That's the way it is with the tubes inside you, too!

Each part of a vascular plant, even the flower, has tubes that transport fluid.

ACTIVITY 1.4
OBSERVE LEAF VEINS

Get a leaf from a nearby plant. Look at it very carefully. Can you see the veins? Do you see one vein in the very middle that is thicker than the rest? That's called the **midrib**. It gets water from the stem and carries it to all of the smaller veins in the leaf. Did you know there are even more veins under the green part of the leaf? They are hard to see right now. But perhaps you've walked through nature and have seen leaves that have lost most of the green and are left with just the veins. If you haven't, begin looking for these leaves when you walk around outside. If you find one, be sure to take it home and put it in your nature journal.

NONVASCULAR PLANTS

Dad

The truth is, most plants have veins, so most plants are vascular. The most obvious way you can tell if a plant is vascular is by looking to see if the plant has roots, stems, and leaves. If so, it is always a vascular plant. Can you think of any plant that doesn't have roots, stems, and leaves? It's hard to imagine plants without roots, stems, and leaves, but they do exist! One that you've probably seen is moss. Moss grows where it is moist, such as the north side of trees or on rocks and land near rivers and streams. Moss, as you've probably already figured out, is a nonvascular plant.

Nonvascular plants don't have inside tubes to transport water through the plant. Instead, they simply absorb water and spread it around as much of the plant as they can the way a paper towel does. As you might have guessed, nonvascular plants need a lot of water around them to live. They can't store water inside themselves the way vascular plants do. Let's look at how nonvascular plants distribute water throughout the plant.

ACTIVITY 1.5a
OBSERVING ABSORPTION

Hart

Here's an activity to help you understand how nonvascular plants get the water they need to survive. You will need a paper towel and some water. Do you think a paper towel has veins or tubes inside it to transport water? No, it doesn't. It absorbs water. That's how it gets wet all over.

Spill some water on a counter and place the edge of the paper

Like nonvascular plants, paper towels absorb water.

towel in the water. Notice how the water spreads through the paper towel. If there is enough water, a lot of the paper towel will get wet. If not, only a small part of it will get wet. That's similar to the way moss and other nonvascular plants get the water they need. There must be a lot of water present in order for water to spread throughout the whole plant. If they don't get enough water, they will dry up. Places that have a lot of humidity (which is water vapor floating in the air) have a lot of nonvascular plants. Why? Because if there is a lot of moisture present in the air, the plants can absorb the water through their leaves. Record what you observed in your Botany Notebooking Journal.

ACTIVITY 1.5b

WALKING WATER WITHOUT A VASCULAR SYSTEM

You will need:

- 4 standard-sized paper towels
- Red, blue, and yellow food coloring
- 5 small, clear cups
- Spoon
- Water

You will do:

1. Line up the cups next to one another and fill every other cup with an equal amount of water.
2. Place a teaspoon of red in the left cup with water, yellow in the middle cup with water, and blue in the right cup with water.
3. Stir the cups of water with the spoon to distribute the coloring.
4. Fold one of the paper towels in half (lengthwise), and repeat two more times.
5. Repeat step 4 with the four other paper towels.
6. Place one end of each paper towel in one cup and the other end in the cup next to it.
7. Check back in a few hours to see how the water has walked.
8. Record what you did and learned in your Botany Notebooking Journal.

Before doing the next activity, explain what you've learned so far about botany.

ACTIVITY 1.6
GO ON A NATURE HUNT

There are not very many nonvascular organisms to be found, but you are going to hunt for them and draw them in your nature journal. Look carefully at the pictures below of the main kinds of nonvascular organisms you may find.

Lichen

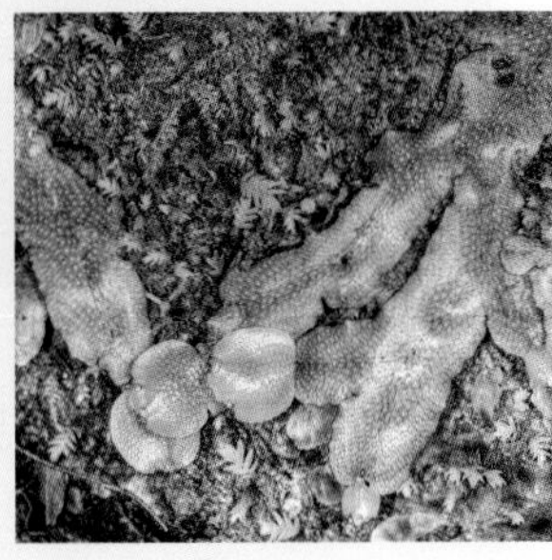
Liverwort

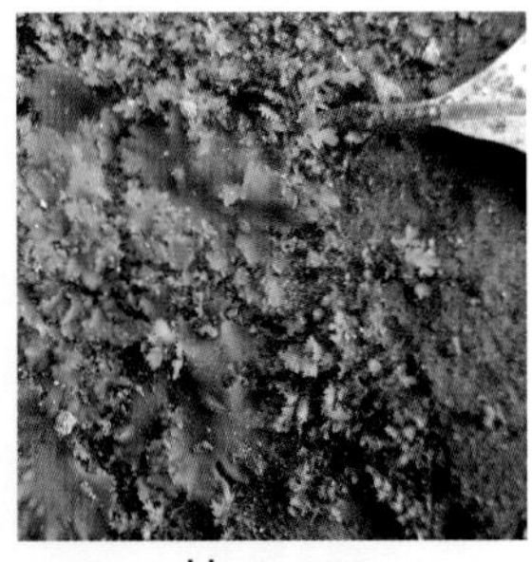
Hornwort

Moss

Now, go outside and see if you can find any of the organisms pictured above. If you find one, use a magnifying glass to look at it closely. You may be surprised that it appears to have a stem or little leaves. But they are not true leaves because they don't have tubes inside. Draw in your nature journal the nonvascular plants you found on your hunt.

TO SEED OR NOT TO SEED

So far you learned that biologists separate all living things into different kingdoms. Do you remember the scientific name for the plant kingdom? You got it—Kingdom Plantae. You also learned that botanists separate plants by whether or not they have tubes inside. Did you know biologists also look at whether or not a plant has seeds? Some plants produce seeds. Other plants do not.

Here's a bit of Latin for you: seed-making plants are called **Spermatophyta**. That's because in Latin the word *sperm* means seed and the word *phyta* means plant. So *Spermatophyta* means seed plants. Biologists divide spermatophytes into even more groups, such as those that make pinecones and those that make flowers. We'll discuss those later in this book.

Stop and think about seeds for a minute. What exactly are seeds good for? Why does a plant

need to make seeds? If you guessed to make new plants, you are right! Seeds grow into a new plant. You would think that all plants make seeds but that's not the way it is. There is a group of plants called spore plants. Plants in this group make spores instead of seeds.

SEEDLESS PLANTS

Have you ever seen a plant called a fern? At certain times of the year, the backside of the fern leaf will be covered with little brown clumps. These clumps are called **sporangia** (spuh ran' jee uh). Inside the sporangia are millions of tiny little bodies called spores. These spores, in the right conditions, can one day grow into a new plant.

Each of the clumps on this fern contains millions of spores.

Even though seeds and spores both grow into plants, spores are not seeds. You see, a seed is a very special plant package. It contains a baby plant, food for the baby plant, and a protective covering. You can think of a seed as a baby plant in a box with its lunch. A spore is just the baby plant and a protective coating. There is no food for the baby plant. Since there is no food in a spore, spores are much smaller than seeds. They also need extra special conditions to grow. We'll learn all about that in a later lesson.

These moss spores are tiny because they don't contain food for the baby plant.

Before you move on to the notebooking activities, let's review what you've learned so far.

WHAT DO YOU REMEMBER?

Why do scientists use Latin and Greek words to name things? What is a biologist? What is a botanist? What are some helpful or interesting things that botanists do? What do vascular plants have that nonvascular plants do not have? What is a Spermatophyta? Can you name one? Can you name a plant that produces spores?

ACTIVITY 1.7
GROUPING PLANTS

In your Botany Notebooking Journal, you will find a special page to help you remember how plants are divided. In each circle, draw or paste a picture of some of the plants found in each group. Below each circle, record what features plants in that group have.

You have already learned a great deal of botany so far. Now it's time to build our special plant growing structure called a light hut. Then we'll start growing our own edible plants!

ACTIVITY 1.8
MAKE A LIGHT HUT

You will need:

- Large, empty, open cardboard box (see image)
- Aluminum foil
- Single socket pendant lamp cord (for lanterns)
- LED full spectrum grow light bulb
- Glue
- Scissors

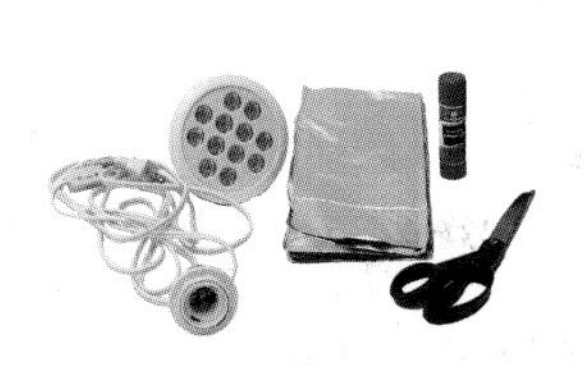

You will do:

1. Cut a 1-inch hole in the top center of the box.
2. Cut ventilation slots in the top, upper sides, and back of the box to allow the air to flow and the heat to escape.
3. Use glue to cover the entire inside of the box with aluminum foil. This will make the inside of the box very shiny.
4. Cut a hole through the aluminum foil where the hole in the top of the box is and cut slits in the foil where the ventilation slots are.
5. Position the light bulb inside the center of the box, pushing the base of the light bulb through the top of the box.
6. Secure the light by attaching the socket from the outside of the box.
7. Tape an aluminum foil curtain to the top front edge of the box so that it hangs down over the opening of the box. This curtain is designed to keep light from escaping the box. However, it needs to hang loosely so there is plenty of ventilation. If your plants seem to be drying out or if the inside of the box gets hot, you will want to lift the curtain for a bit.

ACTIVITY 1.9
GROW EDIBLE PLANTS

If you are beginning your seeds in the fall, grow herbs to keep indoors for the winter. Herbs are a great way to enjoy the plants you grow. Three months before the last frost (you can go online to find out when the last frost is for your area), using the same supplies, pick the seeds you will plant in your edible garden.

You will need:

- Seeds
- Small flowerpots
- Vermiculite
- Peat moss
- Compost
- Your light hut
- Water
- Timer (optional)

You will do:

1. Fill each small pot with equal parts vermiculite, peat moss, and compost.
2. Following the instructions on your seed packet, plant a few seeds in each pot.
3. Water the pots well.
4. Place the pots under the grow light.
5. Keep the light on for 12–14 hours a day, turning it on in morning and off in the evening. You can also put the light on a timer. Make sure you check your box regularly to ensure it isn't too hot or too dry.
6. Water your seeds daily to ensure the soil does not dry out.
7. Check the seed packet to determine the number of days it will take for your seeds to sprout. Make note of this in your notebooking journal. Be sure to note if the seeds sprout on time or if the light hut speeds their sprouting.
8. When the plants begin growing, thin them out so you have only one or two plants in each pot, or follow the seed packet instructions.

Once the seeds have sprouted into seedlings and get too big for the original pots, transplant them into bigger pots that contain equal parts vermiculite, compost, and peat moss. Keep the plants on a sunny windowsill and enjoy the herbs in your cooking!

You can also use your light hut to start fruits and vegetables from seeds. You can then plant them in your edible garden.

LESSON 2

SEEDS

digging deeper

"If you have faith the size of a mustard seed," the Lord said, "you can say to this mulberry tree, 'Be uprooted and planted in the sea,' and it will obey you."
Luke 17:6

How much faith do you have? Do you know that even if your faith is very small, you are still able to accomplish much when you pray to the Lord? Don't let your doubts keep you from praying. It only takes a little faith to do great things!

SLEEPING SEEDS

Before you begin reading this lesson, look ahead to Activity 2.2 and start this activity now by soaking your seeds in a bowl of water. We'll examine them later.

Did you know that inside every seed is a tiny living thing? That living thing is a little baby plant, fast asleep. A baby plant doesn't sleep the way you and I do. It's a different kind of sleep. We say that the plant is dormant. The word **dormant** comes from the Latin word *dormire*, which means to sleep. The baby plant stays dormant until it gets what it needs to wake up. When it wakes up, it begins to grow into a little plant called a **seedling**. When it does this, it uses the soft fleshy material inside the seed for food until it's ready to do what only plants can do. What is it that only plants can do? What's the one thing that separates plants from all the other living things God created? Plants are the only living things that make their own food! Imagine if you could make all the food you need to eat using only three main ingredients. That's exactly what plants do! The ingredients they use are sunlight, water, and air. You'll learn a lot more about how plants make their own food in Lesson 6. In the meantime, let's learn more about those sleeping baby plants, called seeds.

This baby acorn plant has woken up to become a seedling.

What do you think the baby plant needs to wake up? Make some guesses. You were probably really close in your guesses. There are really only three things the baby plant needs to wake up: warmth, water, and air. You might think that a seed needs soil, but it doesn't. The plant will not use the chemicals found in the soil for a few days or weeks after it wakes up. That's because the seed itself has everything it needs to begin growing right inside it. You might think that a baby plant needs sunlight to wake up and begin growing, but it doesn't. Think about this: many seeds begin growing underground where there is no sunlight at all. So there you have it. A seed only needs warmth, water, and air to wake up from its deep sleep and grow into a new plant. Later on, the plant will begin to need sunlight. Sunlight helps the little plant to grow healthy and strong. If there is no sunlight present, the plant will grow anyway, but it will grow differently. It won't be as healthy. Its stem will grow longer and longer, searching for sunlight. If it never finds the sunlight, it will eventually die because it needs light to make food for itself once the food in the seed runs out.

As I told you, before a plant needs soil and sunlight, it needs warmth, water, and air. If the temperature is not warm enough or there is no water or air, the baby plant will stay snuggled inside its seed, fast asleep. But when the temperature is warm enough and water finds its way into the seed, the baby plant begins to emerge from inside the seed. If not, it'll stay fast asleep until the time is right.

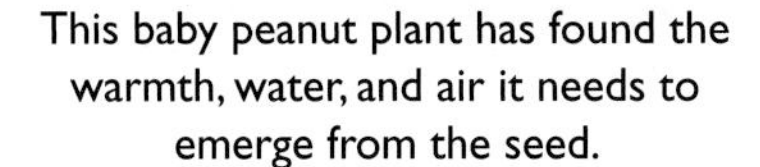

This baby peanut plant has found the warmth, water, and air it needs to emerge from the seed.

So how long can a seed sleep before it can grow into a plant? Sometimes a baby plant can sleep for years. Believe it or not, a few kinds of seeds can sleep for thousands of years. Now that's a long sleep, isn't it? Most seeds, however, cannot grow if they aren't awoken within two to eight years. When a plant makes a seed, it usually finds itself in the right conditions to begin growing as soon as it hits the ground. That's the wonderful way God created plants to make new plants.

Let's take a closer look at seeds. As you know already, inside every seed is a tiny baby plant. That baby plant is called an embryo (em' bree oh). Did you know that you were once an **embryo** when you were inside your mother's womb? Of course, you weren't a plant embryo, and you didn't act like one either! You weren't sleeping all the time. No, you weren't dormant at all. Instead, you were wide awake, moving, jumping, kicking, and sometimes sleeping. As you know, God created humans and plants very differently. However, if you think about it, you can see the signs of the same Creator in both humans and plants. We all begin as embryos, and we all begin protected inside something. Your mother's womb protected you when you were an embryo. What protects a plant embryo? Can you guess? Did you guess the shell of the seed? That's right, and now it's time for you to learn more about a seed's shell!

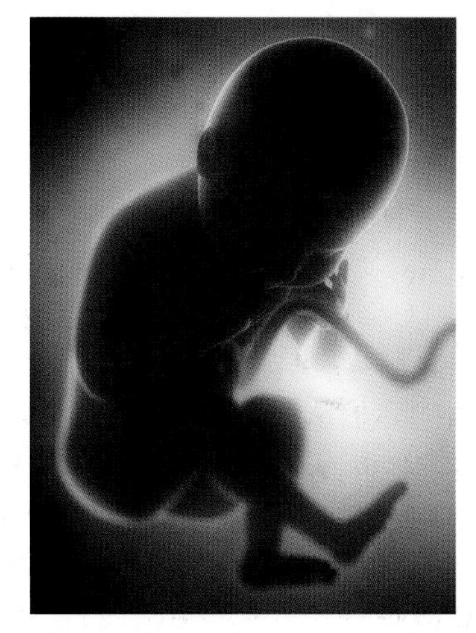

Like the plant embryo protected in its seed, you were protected as an embryo in your mother's womb.

TESTAE

Gather your seeds that have been soaking in the water and pick them up. Look closely at the shells of your seeds. The shell is sometimes called the seed's coat. The scientific name for the seed's coat is **testa** (test' uh). The testa protects the embryo, much like a winter coat protects a person from a cold winter's chill. Think about that. God created a special coat for each embryo. Let me ask you a question. What color is your winter coat? Wouldn't it be strange if everyone in the world

These children's coats protect them like a plant's testa.

had the exact same coat? That would be kind of boring, wouldn't it? Well, God must like things to be exciting because He gave different coats to every type of seed He made. This means each plant's seed has a different testa. Testae (plural of testa) come in all shapes, sizes, and colors. Seeds from some plants have thin testae, like a spring jacket. Seeds from other plants have testae that are thick and strong, like an Eskimo's coat. The walnut seed's testa is wrinkly and very strong. Corn seed testae are thin, smooth, and yellow. Lima bean testae are thin, smooth, and white. An acorn has a little "hat" to go along with its winter coat. Have you ever seen a coconut? It's a large, furry seed that grows on a coconut palm tree. Think about the seeds you planted. How were they different from one another?

This furry testa belongs to a coconut seed.

The acorn seed's testa provides a coat and a hat as well!

God made the walnut seed's testa wrinkly and strong.

So the testa is there to protect the seed like your coat protects you. But what do you do if your coat gets wet or it becomes warm outside? You take it off! That's what the seed does when its coat gets wet and warm. When the seed is exposed to water, such as after a warm spring rain, the seed absorbs water through its seed coat. This is called **imbibition**. This makes the testa soft and soggy. The testa then comes off, and the sleeping embryo continues absorbing the water. The bean seed that's been soaking in water is probably not as firm as it was before you put it in the water. Its testa is trying to come off!

It's important for you to understand that what you are looking at is called a mature seed. That means the seed can sprout as soon as it gets the water, warmth, and air it needs. When the seed is still attached to its mother plant, it is not yet mature. Look closely at the bean testa with a magnifying glass. In the curve of the bean seed there is what looks like a little scar. Believe it or not, that's the seed's belly button! Every seed has a belly button where it was attached to its mother. It's actually called a **hilum** (high' lum). You have a belly button as well, and it's also the place where you were attached to your mother. So, you see, God designed so many things in a similar way.

Can you find the hilum on the sunflower seed? It's not on the pointed end; it's on the other end of the sunflower seed. If you have other seeds, you can try to locate the hilum on each one.

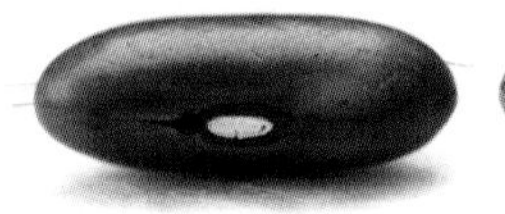

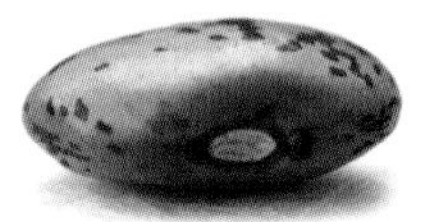

ACTIVITY 2.1

DESIGN A SEED

With a parent present, search the internet for unusual seeds. After you've examined the many fascinating seed coats God created, design a special seed of your own creation in your Botany Notebooking Journal. Also, draw the kind of plant that your seed would grow into.

ACTIVITY 2.2
EXAMINE YOUR SEEDS

Would you like to see a plant embryo for yourself? Soak some seeds like a bean and a sunflower seed in some water. If your seed has been soaking long enough (a few hours), you can gently pull the testa off. Then, very, very slowly and very, very carefully pull the seed apart. If you were careful, you should see a tiny growth inside the seed. Use a magnifying glass to look closely. Doesn't it almost look like a tiny plant? That is part of the embryo. Study it more with your magnifying glass. Add what you've learned to your Botany Notebooking Journal. After that, read on to learn about each part you see.

ANATOMY OF A SEED

There are five main parts to the embryo in your seed: the **radicle** (rad' ih cul), the **hypocotyl** (hi' puh kot' uhl), the **epicotyl** (ep' uh kot' uhl), the **plumule** (ploom' yool), and the **cotyledons** (kot' uh lee' dunz). Can you find and point to these parts on your embryo? Now that you know the names of the five main parts of a seed, let's learn a little bit about what these parts do.

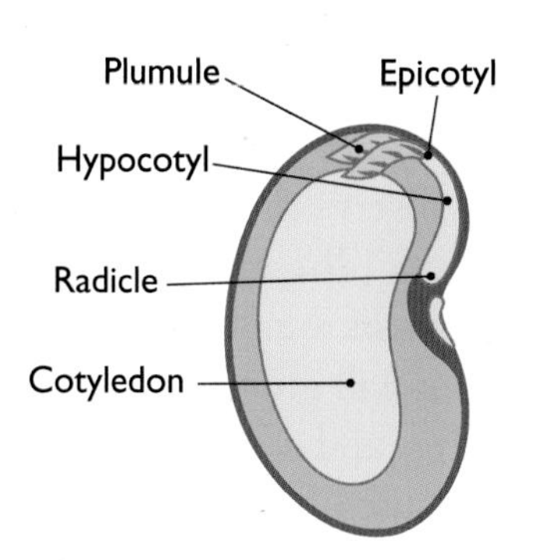

INSIDE THE SEED

The **cotyledons** are the biggest part of the embryo, and they have an important purpose. They provide food for the embryo once the seed opens and the embryo starts growing into a plant. Your bean seed has two cotyledons. Some seeds, like corn, only have one. If a seed has two cotyledons, it's called a dicotyledon, or **dicot**. Di means two. This tells us that dicots have two cotyledons. Mono means one. There is only one cotyledon in a **monocot** seed. Monocots and dicots produce plants that are slightly different. We'll discover more about that in a moment. The most important thing to remember now is that the cotyledon provides the plant with the food it needs to grow. The food the plant eats is called the **endosperm** (en' doh spurm).

When the bean seed is immature, its cotyledons are small, and it has an endosperm. Endo means inside and sperm means seed, so endosperm means inside the seed. The cotyledons absorb the endosperm food. This causes the cotyledons to grow into the biggest part of the embryo. When the seed finally opens and the embryo begins to grow, the cotyledons give the new plant the endosperm food they have absorbed. Think about this: when we eat beans, we consume the endosperm from the cotyledons—the food meant for the baby plant!

Leaf formed from the plumule

Section formed from the epicotyl

Cotyledons

Section formed from the hypocotyl

Now remember, God made a lot of different kinds of seeds. Not all of them look like bean seeds. In some seeds, the cotyledons don't absorb the endosperm before the seed is mature. Instead, they start to absorb the endosperm after the seed has opened. Then, they pass the food immediately to the embryo. In those kinds of seeds, the cotyledons are much smaller than in the seed you've been looking at, and the endosperm is there as a separate part of the seed.

Now that we've discussed the embryo's food, let's examine the tiny parts of your embryo that will one day grow into a plant.

Look for the **radicle**. The radicle is the embryo's root and grows into the plant's root. When the radicle grows, it makes tiny little hairs. Those hairs are not hairs like you and I have; they are little fibers that go searching for water and nutrients. When they find these things, they absorb them and send them to the rest of the plant. In this lesson, you'll do an experiment where you get to watch the radicle grow and see these tiny little hairs.

The **hypocotyl** will become the stem. If light is present, it won't need to grow very long. However, if light is not present, the hypocotyl will elongate (grow longer), in search of light. You'll do an experiment in which you deprive a few seeds of light, while others receive light. It will be interesting to observe the differences in how they grow.

The **epicotyl** is the top of the embryo. It holds the **plumule**, which becomes the first true leaves of the plant. Why do I say "true" leaves? Because when your embryo begins to grow, the cotyledons will grow upward on the hypocotyl and actually look like leaves. They're not as green as the true leaves and aren't shaped the way the plant's leaves will be shaped, but they look very similar to leaves. Because of this, some call the cotyledons the first leaves of the plant. We often call them **seed leaves** because they are really part of the seed. However, remember that they are not true leaves. The plumule will be the first *true* leaves. **Plumule** is a Latin word that means feather. The plumule of your bean seed sort of looks like feathers, doesn't it?

Keep in mind that if you open other seeds, you may not see a plumule. Most embryos don't have their true leaves in this stage. Have you ever seen a baby born with teeth? It doesn't happen often, but when it does, it's a big surprise. Usually the teeth don't grow in until a lot later. That's what it's like for the bean plant. It already has its first true leaves, even though most plant embryos do not.

Since the bean plant starts out with its true leaves already developed, it has a head start on growing. After all, the leaves are a vital part of the plant. They actually make food for the plant. Isn't it amazing how unique plants can be? Some have a head start on growing, others take more time to mature. You'll soon discover that some plants grow best in warm weather and others grow best in cold weather. God designed our world to give us food in many different climates and seasons.

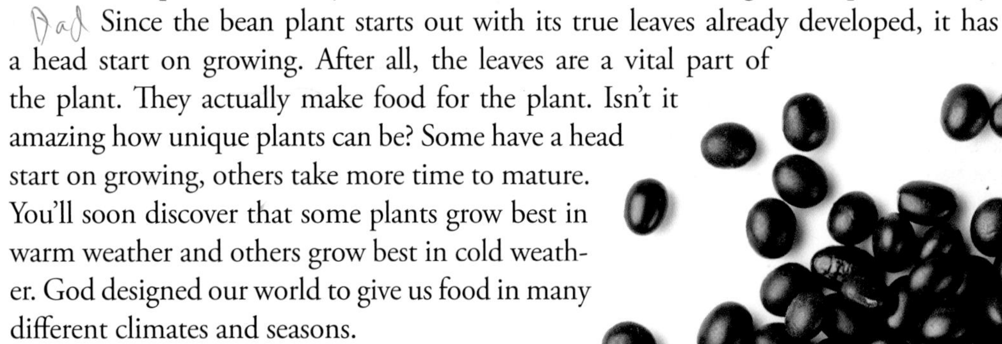

Tell someone everything you have learned so far about seeds.

MONOCOTYLEDONS AND DICOTYLEDONS

Do you remember that scientists like to put plants into different groups with similar features? Well, botanists classify some plants based on the number of cotyledons the seed has. Do you remember I mentioned monocots and dicots? They're two groups of plants, and after reading this section you'll be able to go out in your yard and classify the plants you see as monocots or dicots.

It's time to put on your scientific observation hat. Look carefully at the four images below. How are they different from each other?

Are you ready to learn the difference? It's pretty easy. You see, monocot leaves have veins going upward from the bottom to the top of the leaf rather than branching out from a thick, center midrib vein. So when you're out identifying plants, look first at the way the leaf veins are organized. If the veins never branch out but instead form lines that run up and down the leaf, the plant is a monocot. If the veins branch out from a thick vein running down the middle of the leaf, the plant is a dicot.

Another difference between monocots and dicots can be seen if you look at their flowers. Monocot flowers usually have petals in multiples of three. Can you count by threes? Three, six, nine. If you count the petals on a flower and find that there are three, six, or nine petals, then the plant is most likely a monocot (unless a petal fell off). On the other hand, dicots usually have flowers with petals in multiples of four or five. In other words, dicot flowers might have four petals or eight petals. A chrysanthemum has too many to count! That's a lot of petals. Many flowers have so many petals that it's much easier to identify them by their leaves.

Because they have many petals, it's much easier to identify these chrysanthemums as dicots by their leaves.

ACTIVITY 2.3

IDENTIFY DICOTS AND MONOCOTS

Find a place outside or go to a plant nursery that has a lot of flowers and plants. See if you can identify whether they are monocots or dicots. Take your nature journal along and record what you observe.

GERMINATION

So far, you've learned that when a seed gets wet, the testa loosens. Water then gets inside the seed and the embryo wakes up! Did you know that this simple action starts a special activity called **germination** (jur' muh nay' shun)? Germination is simply the process of a seed sprouting into a plant.

This image shows the growth process of a bean becoming a plant.

What happens during the germination process? First, the radicle pushes its way out of the seed. Do you remember what the radicle is? It's the seed's roots. It makes sense that the roots are the first things out because the roots are what take up the water the plant needs to grow. As you can imagine, the roots typically grow downward.

After the roots emerge, the hypocotyl and epicotyl move up and out. They must grow straight up because they need to break through the dirt to reach the surface of the soil. Look at the image on this page to see how the plant looks under the ground as it's growing.

Once the hypocotyl and epicotyl break through the surface, the hypocotyl becomes the stem, and the epicotyl holds what will become the first true leaves of the plant (the plumule).

For the bean seed and other seeds that grow a large cotyledon, you will see the cotyledon between the epicotyl and plumule. Remember, the cotyledons look a lot like leaves, but they are really only seed leaves.

All the nutrition it will need for several days is packed within the seed's endosperm or cotyledon, depending on the seed.

Can you guess what kind of food the endosperm is filled with? What is your very favorite thing to eat? Ice cream? Cake? Cookies? Whatever it is, it just might have some sugar in it. Guess what? The endosperm also has sugar in it. It has other nutrients the plant needs as well, but sugar is a main ingredient. Plants

Plumule (attached to the epicotyl) forms the first true leaves
Cotyledons supply food to the developing plant and become what some call the "seed leaves"
Testa comes off
Hypocotyl develops into the stem
Radicle emerges from the seed
Radicle develops into the root

People are consumers, depending on plants and animals for food to survive.

love sugar! They love sugar even more than you do because they cannot live without sugar. It's their food. After plants take all the nutrients from the endosperm, they begin to make their own sugar.

Did you know that we (and most of the creatures in creation) depend on other living things to survive? We consume other parts of God's creation, like plants and animals, to live. So we are called **consumers**. Since God created plants to make their own food, they don't eat other plants to survive the way we do. They also don't eat animals to survive. However, there are some plants that do trap and absorb animals, but they are very rare. We'll discuss those strange plants in the next lesson. Nevertheless, it's important to remember that plants produce their own food. Because of this they are called **producers**.

Before we end this lesson and move on, let recall what you've learned by answering some questions. It's okay if you need to look back in your book to find the answers. When you are done, we have some fun activities to end our lesson.

WHAT DO YOU REMEMBER?

What is a seed? What does dormant mean? What does a seed need to wake up and begin growing? What is the baby plant in a seed called? What is the seed's testa? What does the testa do? What is the hilum on a seed? Describe germination. What is the top part of the embryo called? What are the featherlike leaves on the embryo called? What is the embryo's root called? What is the nutrition within the seed called before it gets absorbed by the cotyledons? Explain how the testa comes off for germination. What is a producer? What is a consumer? Are plants producers or consumers? Are people producers or consumers?

ACTIVITY 2.4

LABEL THE PARTS OF A SEED

Label the parts of a seed in your Botany Notebooking Journal. Also, write down everything you can remember and make illustrations about what you have learned.

ACTIVITY 2.5
COMPARE GERMINATION CONDITIONS

Let's grow some seeds inside plastic bags so that you can watch their development and make a guess about which ones will grow best. You'll write down all you're going to do for this experiment and make a good guess about what you think is going to happen on the special science experiment sheet provided in your Botany Notebooking Journal. You can also find a copy on the Book Extras website for this title at Apologia.com.

You will need:
- 3 plastic Ziploc® bags
- 3 paper towels
- 3 or more turnip seeds (or bean seeds)
- Tape
- Ruler (should read centimeters)
- Scientific Speculation Sheet

You will do:
1. Wet three paper towels with water and place one inside each of the plastic bags.
2. Place a seed in each bag. If you have more than three seeds, you can put more than one seed in each bag. Zip each bag closed once you have put the seed or seeds in it.
3. Tape the first bag to a window that receives sunlight. Make sure your seed faces the window.
4. Place the second bag in your refrigerator.
5. Place the third bag in a dark closet that no one goes in and that never gets any light.
6. On your Scientific Speculation Sheet, write down your experiment and your guess about what will happen to each seed. For example, guess whether or not each seed will germinate, and guess which one will grow fastest and which will grow slowest. This is called a **hypothesis** (hi pahth' uh sis). When you write down your hypothesis, also include why you made that guess.
7. Make a Seed Growth Chart for each bag or use the one found in your Botany Notebooking Journal. In other words, you will have one Seed Growth Chart for the seed that is in the bag taped to the window, another Seed Growth Chart for the seed in the dark closet, and another for the seed in the refrigerator. A sample Seed Growth Chart is shown on the next page.
8. To fill out each chart, use the ruler to measure the length of the plant (not the seed) each day. Put a dot in the square that marks the day you are making the measurement and the measurement that you made. For example, today is Day 1. Your seed has not grown at all today, so put a dot in the square that has "Day 1" on the bottom and "0 cm" to the left. Tomorrow will be Day 2. If any of the seedlings have sprouted, measure and record their lengths.

6 cm												
5 cm												
4 cm												
3 cm												
2 cm												
1 cm												
0 cm												
	Day 1	Day 2	Day 3	Day 4	Day 5	Day 6	Day 7	Day 8	Day 9	Day 10	Day 11	Day 12

9. At the end of 12 days, compare the Seed Growth Charts. The seed that has dots highest on the chart the soonest is the one that grew the fastest. Which one was that? Was your hypothesis correct? Record the results of your experiment in your Botany Notebooking Journal.

LESSON 3

ANGIOSPERMS

digging deeper

Consider how the wildflowers grow: They don't labor or spin thread. Yet I tell you, not even Solomon in all his splendor was adorned like one of these. If that's how God clothes the grass, which is in the field today and is thrown into the furnace tomorrow, how much more will he do for you—you of little faith?
Luke 12:27–28

God filled Earth with exquisite color when He created flowers. Flowers come in an endless variety of shapes, colors, and smells. They are beautiful and unique, and each one makes the world a more interesting place. You are like a flower in this way because you are not like any other person God made. You are unique and special and make the world a more wonderful place. It's no wonder God cares so much for you!

FLOWERING PLANTS

Hal Now that you've learned all about seeds, it's time to focus on where seeds come from. Seeds come from a kind of plant called an angiosperm. Angio means container and sperm means seed. All angiosperm plants make some kind of container for seeds to grow in.

Harleigh You might be thinking: Plants make containers? That's weird. What kind of containers? Well, that depends. We don't usually think of fruit as a container, but that's exactly what it is: a container to contain seeds! Some plants make a prickly sticker for their seeds. Others make a container with a furry parachute attached so the seeds can go flying in the wind. God created many different kinds of containers for angiosperm seeds. Can you think of any seed containers that you have seen? Did you know that an apple is a container for apple seeds?

Harlan Did you know that before the apple, pea pod, or acorn shows up on the plant there's a flower there first? All angiosperms make flowers before they make their seeds. The best way to figure out if a plant is an angiosperm is to ask, "Does it make flowers?" If you can answer "yes," then the plant is an angiosperm.

Dad Angiosperms are very important plants. That's because they grow most of the food we eat, like wheat, apples, potatoes, and rice. And without angiosperms, we would not have cotton plants to make our clothes and sheets. There would be no sugar cane plants to give us sugar and fill

Apple seeds come with their own containers.

This cotton flower shows us that the plant is an angiosperm.

our birthday parties with cakes and cookies. There would be no maple trees for the sweet maple syrup that drips on our pancakes. In fact, no angiosperms would also mean no pancakes! Not only that, angiosperms are an important source of food for birds and mammals. Without flowering plants, many animals would die.

In fact, if there were no flowers, there would be no nectar for bees. The bees use the nectar in flowers to make honey. No flowers. No bees. No honey. The world without angiosperms is already sounding a lot less sweet, isn't it?

We can be thankful God created angiosperms. Not only do they provide our lives with essential things we need to survive on Earth, they are often quite beautiful. Have you ever been to a botanical garden? There you will see flowers of every color, shape, and size. Flowers truly make the world a more beautiful place. People have enjoyed flowers since the Garden of Eden. In fact, one of the Seven Wonders of the Ancient World is a garden: the Hanging Gardens of Babylon. This amazing garden is believed to have been built by King Nebuchadnezzar II long ago, about five hundred years before Jesus was born. What made this garden a wonder of the world is the way the gardeners engineered the water to keep this lush, green garden alive in the hot desert of what is now Iraq.

This artwork depicts what the Hanging Gardens of Babylon may have looked like.

Later in this book, you'll learn how to grow a fruit and vegetable garden. You can then use your new gardening skills to grow a beautiful flower garden as well.

The fact that God created so many different kinds of beautiful flowers for us to enjoy is quite remarkable. Did God have to make the flowering plants so lovely for our eyes to behold? No. The Lord did not have to make the grass on our lawns such a beautiful, calming green color, or the trees so majestic and tall against the sky, or the flowers in spring so full of splendor and beauty. He could have just as easily made all the trees filled with leaves of a dull gray color, the grass the same dull gray, and the flowers all the same dreary, uninteresting gray. He did not have to make all the amazing hues of pinks, blues, purples, yellows, oranges, and reds that we see. But He did. All these beautiful colors are a feast for our eyes and a gift from the Lord.

God's creative beauty is seen in the colorful flowers He made.

Just like when you create something, God took great care and delight in everything He made.

Why do you think God made the world so much lovelier and more interesting than He had to? Have you ever taken a long time to draw a picture? Did you put a lot of thought and detail into the picture? If so, then you have a glimpse into how God feels about His world and all He created. He took great care and put much thought into everything He made. The beauty God placed in the world tells us about the character of God. We know that He cares about us and wanted to make Earth a lovely place for us to live. When you enjoy the sunset, the flowers, the trees, and a nice cool breeze, remember they are gifts from God and remember to thank Him in your heart.

The things we draw or work on are special to us while we are focused on them. Sadly, we easily forget about the things we've created or done because we are always moving on to the next thing on our list. Thankfully, God never does that. He is always focused on His creation, diligently working on the details of every single thing that happens on Earth even though He formed it long ago. God sees every caterpillar egg that hatches, every hair on your head, and every tear in your eye. He loves you so much and wants the best for you. He is always making sure that things work out according to His perfect plan for your life. All of nature must obey Him, and He is in control of all things.

Just like everything else, flowers obey God's plan and design for them. God's purpose in creating them was not only for our needs and pleasure but also for a very grand purpose. The flower's job is to make babies! Do you remember where all baby plants begin? They begin as an embryo in a seed. And all flowers make seeds, even the tiniest ones. That is their God-given purpose in life.

God designed flowers in such a special way, and I want you to learn more about them. To do this, we'll take apart a flower. When we take something apart to study it, we say we are **dissecting** (die sekt' ing) it. As you dissect a flower, you'll be learning about flower **anatomy**. Anatomy simply means the study of the different parts of a living thing. The anatomy of the flower will teach you about how flowers make seeds.

Before you dissect a flower, explain what you have learned about angiosperms.

ACTIVITY 3.1

DISSECT A FLOWER

You will need:

- Flower (A lily works best because all parts of the flower are easily visible.)
- Flower dissection page of your Botany Notebooking Journal
- Glue or tape
- Adult to use a knife in this activity

You will do:

1. Study your flower for a moment.
2. Turn your flower upside down. On most flowers you'll see green leaf-like points under the petals. Does your flower have them? If not, they may have fallen off. They're called **sepals** (see' puls). Look at the sepals on the rose pictured to the right. The sepals work together to cover and protect the developing flower. Before the flower blooms, it's a bud. The bud is covered by the sepals, which are usually green. As the bud opens, the sepals separate, ending up under the flower. Because the sepals are no longer needed after the flower blooms, they often fall off. On certain flowers, however, the sepals stay attached to the base of the flower, just above the stem. All of the sepals together are called the **calyx** (kal' ix) of the flower.
3. Using your fingers, carefully pull each sepal off your flower. Once you've removed all the sepals off the calyx, lay them flat on the flower dissection page of your Botany Notebooking Journal and secure them with white school glue. Don't worry, the glue will dry clear and the flower parts will retain their color. Be sure to keep your journal open for a few days so the glue can dry. You don't want your pages to stick together!
4. Using your fingers, carefully begin removing the petals one by one. Be certain to remove each entire petal, not just the uppermost part. The petal reaches all the way down to the top of the stem. All of the petals together are called the **corolla** (kor oh' luh). When you have removed the petals of the corolla and glued them on the page in your Botany Notebooking Journal.
5. Did you get any slimy stuff on you as you removed the corolla? That slimy stuff is the nectar (nek' tur), which is kept at the base of each petal. Birds, bees, butterflies, and many insects love this sweet juice, which is why you see them flying around flowers! Do not eat the nectar. If your hands are sticky, take a moment to wash them.
6. What you now have left are the most interesting parts of the flower—the boy and girl parts! Yes, indeed! Almost all flowers have male and female parts within them. These parts aren't male and female the way you and I are male and female, but they are similar enough that botanists have named them that way. Remember, the same Creator that made you and me also made the flowers. Of course, then, all that He made would be somewhat similar. The male part of a flower makes the female part of a flower pregnant so that it can make seeds. Let's learn more about this exciting seed-making business.

Before we go on, tell someone what you've learned so far about flower anatomy.

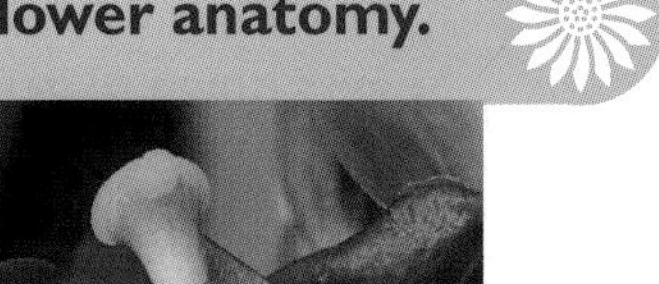

7. The male parts of the flower are the little stalks, or poles, that all look basically the same. Each of these little stalks is called a **stamen** (stay' men). It's easy to remember because males are men, and stamen has the word *men* in it. The stamen's job is to make pollen. You should see one stalk that looks different from the stamen. That is the female part of the plant. Can you recognize the stamens in your flower? You may only have a few compared to what is pictured in the flower on the right.

8. Each stamen has a pole called the **filament** (fill' uh ment). Sometimes the pole is tiny; sometimes it is very long. Sometimes it is attached to a center portion of the flower; sometimes it simply surrounds the center structure of the flower. At the top of every filament is an enlarged part covered with pollen called the **anther** (an' thur). Can you find the anther and filament on one of the stamens in your flower?
9. Remove an anther from one of the stamens. Be careful not to remove the female part of the flower. We'll learn about the female part in a moment. If you have a magnifying glass, use it to study the anther. Shake a bit of the dust off if it's still there. Place the anther on your journal page and glue it down. Anthers are usually covered with thousands of pollen grains that are very small. Combined together these grains make a fine dust called pollen. Inside each pollen grain are two little sperms. Do you remember what a sperm is? We learned from the word angiosperm that sperm means seed. Pollen, then, is a kind of seed! In fact, it's really about half of a seed. When pollen reaches the female part of a flower, it will dig into the female part until it finds a tiny egg there. It will then join with that egg to make a seed. Isn't that simply astonishing?
10. Remove the rest of the stamens from your flower and glue them on your journal page.

Before we go on to study the female part of the flower, explain in your own words what you have learned about the male part of the flower. Make sure you explain what the male part of the flower is called, what its parts are called, and what its parts do.

11. The only thing you should have left of your flower is the female part and the stem. The female part of the flower is called the **carpel** (car' pul), and it's attached to the center of the flower. There is usually only one carpel per flower, but sometimes there are more. How many carpels does your flower have? Most carpels are long and thin, but some are short and fat. Look at the flowers to the right and notice the different shapes that carpels come in. Is your carpel like any of these? Carpel shapes, sizes, and colors are some of the most interesting things to study. God truly used marvelous and imaginative design with all the unusual carpel shapes. You would be simply amazed at all the distinct kinds of carpels there are in this world. At the top of the carpel is a sticky head called the **stigma**. The stigma is sticky because it's designed to catch any pollen that touches it. The pollen then grows a tube that reaches the egg down inside the carpel. The stigma is on top of a long tube called the **style**. This is the tube that the pollen goes down to get to the egg. Down at the bottom of the style, the very bottom part of the carpel is the flower's **ovary** (oh' vuh ree). Each ovary contains tiny egg-shaped structures called **ovules** (oh' vyools).
12. Now that you've removed everything from your flower except the carpel, it's time to dissect the carpel to see if there are ovules inside. Using a knife is very dangerous; please have a grown-up slice your carpel in half vertically as shown in the drawing on the top of the next page. Inside the ovary of your carpel, you might see tiny things that look like eggs. Those are the ovules. The ovules con-

tain the eggs, which are too small for you to see. Do you remember when I told you that pollen is like half of a seed? Well, the eggs inside the ovules are the other half! When the pollen digs down through the style and into the ovary, it will find an ovule and meet up with one of the eggs. When the two join, the plant has everything it needs to make a seed. Isn't that remarkable?

Attach the carpel to your Notebooking Journal. Keep your flower dissection page open until all the glue dries. To help you remember all the parts of a flower, I've included a drawing below. Now that you've seen a real flower and have touched all its parts, you should be able to understand this drawing.

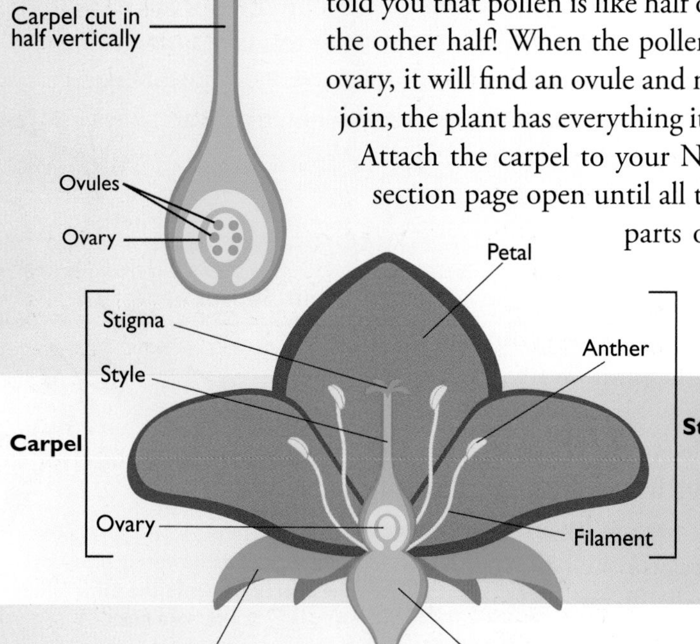

Tell someone about all the parts of the flower you have studied. Be sure to tell where the eggs and pollen are and how the pollen gets to the eggs.

ACTIVITY 3.2

LABEL A FLOWER

In your Botany Notebooking Journal, draw a diagram of a flower, including all the major parts you learned about in this lesson (stamen, carpel, petals, and sepals). Label all the parts of the drawing. After that, record what you learned about flower anatomy.

ACTIVITY 3.3

WALK IN NATURE

Grab your nature journal and supplies and go outside for a nature walk. Begin searching for flowers of all varieties. When you come to a flower, study it closely. Make a sketch of the flower's carpel, stamens, petals, and sepals. Notice the carpel in particular. You will find great variety in the design of carpels as you search from flower to flower. Next to your sketch, illustrate the entire flower and even its plant if there's room. If you know what kind of plant you're drawing, you may want to make note of that. Be sure to record the date, time, and place!

If flowers are not in season at this time, plan a trip to a nursery or botanical garden that has an indoor garden you can observe.

FLOWER FAMILIES

There are so many angiosperms in the world that botanists have created about 300 families of them. In each family, the flowers all look very similar. There is the lily family, the daisy family, the rose family, the mint family, the violet family, and on and on it goes.

How similar are the plants in a single family? Well, let's think about the rose family. It includes roses, but it also includes flowers that grow on some trees, such as the cherry tree. Why are cherry trees grouped in the same family as roses? It's because their flowers have very similar features. Sure, there are a lot of differences between a rose bush and a cherry tree, but they have enough in common to put them into the same family.

Because of its flowering blossoms the cherry tree is in the same family as the rose bush.

DAISY FAMILY

Since there are so many families of angiosperms, I cannot talk about them all. However, I do want to spend some time telling you about a few interesting flower families. One is the daisy family, whose Latin name is *Asteraceae* (as' tur uh see' uh). One of the flowers in the daisy family is called an aster, which is the Latin word for star. This family is interesting because its members produce **composite flowers**. Some composite flowers you may be familiar with are sunflowers, chrysanthemums (or mums), daisies, and asters. Look at each of the flowers pictured and describe something you see that is similar about them all.

It's really difficult to see the stamen and carpels of these flowers. That's because the center of the flower is a mass of tiny structures. It's what makes these flowers composite flowers. The word composite means something made of many parts. Look at the picture below. If you have a dandelion like this growing in your yard, get it so you can study a composite flower in real life.

Like I said before, each composite flower has a special thick mass in the center. That thick mass is actually a group of tiny flowers that are growing upon

The center of this dandelion is a mass of tiny structures, making it a composite flower.

Sunflowers are called ray flowers because their petals surround a central disk like the rays of the sun!

a round, flat disk. What at first seems to be a single large flower is actually a composite, or combination, of many smaller flowers. Each of these teeny, tiny flowers has five itty-bitty petals fused together. And believe it or not, each minuscule flower in the center has its very own tiny stamens and carpel, with an itty-bitty ovule deep within. What's even more interesting is that each of these little flowers produces just one single seed. Arranged around the composite flower are structures that look like petals. But even those "petals" are actually flowers. Each petal is a single flower all by itself. They're called **ray flowers** because they surround the central disk like the rays of the sun. Now that's an interesting flower, wouldn't you say?

Sunflowers are some of the easiest flowers to grow. They're really interesting because the flower twists and turns on the stem all throughout the day so that it always faces the sun. That's why it's called a sunflower! You can plant a sunflower seed in a cup of soil. Keep it moist and warm. It takes about a week for the seed to germinate. Then, it begins to grow taller and taller. Plant it outside in a sunny location and keep it watered. It will die back in the winter and return again in the spring. We call plants that do this **deciduous**.

ACTIVITY 3.4
PLANT A SUNFLOWER

You will need:
- Cup of soil
- Sunny windowsill
- Sunflower seed

You will do:
1. Using a pencil or your finger, poke a 1-inch hole in your soil and drop the sunflower seed inside. Cover the seed up with soil.
2. Place the cup with the seed on a sunny windowsill.
3. Keep the seed watered and wait until the sunflower has grown.
4. Plant the sunflower in a sunny location if it's warm outside.
5. Watch the sunflower turn to face the sun as the sun travels across the sky from its rising in the east to its setting in the west.

ORCHID FAMILY

The orchid family is one of the largest flower families on Earth, with more than 30,000 different species, or varieties, grown in the wild. Even today, new species of orchids are being discovered. Orchids are so strange and beautiful that they've captured the hearts of people for thousands of years. In fact,

This unusual orchid is one of 30,000 orchid species growing in the wild.

God used great variety and ingenuity when designing orchids.

This lady slipper orchid resembles a ballet slipper, don't you think?

Do you see how this orchid looks like a bee? It smells like one too!

you can find groups and events completely dedicated to orchids.

Most native orchids live in the tropical rainforest, often growing in the nooks of trees, sometimes hundreds of feet in the air. Orchids that grow on trees are called **epiphytes**. Epi means above and phytes means plants. So orchids are above plants. Some people call them air plants. However, there are so many varieties that some orchids can grow in any environment, including in deserts, on the ground, under the ground, anywhere! A few kinds of orchids are fairly easy to grow, while others are so rare and delicate they require as much care as a newborn baby. An orchid plant can be as small as a penny or as big as a compact car. Orchids that live in desert regions have thick leaves, almost like a cactus does, while those that live in the tropics have long, thin leaves. Some orchids have no leaves at all!

Look at the pictures of orchids. One is a lady slipper orchid. It has a deep pocket that's actually a bee trap. When a bee crawls down into the flower to get the nectar, the pocket on the flower closes shut. As a result, the bee is stuck for a while, wiggling and squirming, getting pollen all over itself. The bee eventually finds a small opening near the top of the orchid and escapes so it can get trapped inside another lady slipper, where it will transfer the pollen to the new lady slipper's stigma. As you know, that's called pollination.

Study the image of the bee orchid. Doesn't it look a lot like a bumblebee? Not only does it look like a bumblebee, but God also designed this flower to actually smell just like a female bumblebee! This smell attracts male bumblebees, which come to the flower for a friendly meeting. Before the male bumblebee realizes he has been tricked, the orchid gets pollinated! A few other orchids look and smell like insects in order to attract pollinators, and many have traps like the lady slipper.

One kind of orchid stinks like the rotten food a fly or gnat would want, attracting hordes of flies and gnats to its smelly nectaries. These flies and gnats search around for food, pollinating the flower in the process. The insects then leave because they cannot find the meal they desire. You see, they don't want the orchid's nectar. They want rotten food, but there is none in these orchids even though it smells like it. If you smell a flower that simply stinks, you know it was designed for the flies and gnats of the world to pollinate.

THINK ABOUT THIS

Flowers don't use nectar for themselves. They only use it to attract animals to help them in pollination. Flowers spend a lot of energy making nectar that just gets eaten by the animals. Since orchids like the bee orchid get pollinated without actually feeding the animals, survival is easier for them. They don't have to keep making food for animals. Evolution would say that since these orchids have an easier time surviving than orchids that actually feed animals, they should be the main kind of orchids in creation. Why, then, are they rare compared to the other orchids? Only a few orchids attract animals by imitation. Most of them use a lot of energy making food for animals in order to attract them. The fact that most orchids (and flowers in general) produce nectar for animals to eat shows that God intended flowers and animals to work together to survive.

With their leaf shapes, you probably guessed that orchids are monocots. But orchids reproduce seeds differently than most other monocots.

You see, when an orchid flower makes seeds, it doesn't just make one like each tiny petal of a sunflower. It makes millions of them! And they are so tiny that they look like nothing but dust. Orchids have the smallest seeds of all the angiosperms. Sadly, only a few of the seeds will actually become another orchid plant. That's because orchids' seeds are one of the few seeds that don't have an endosperm at all. The seed must find its way to a certain kind of living matter that can provide food for the embryo. The new seeds blow around until they land somewhere. Those that land near a certain kind of fungus are the lucky ones. The fungus helps the seed grow into a plant by providing nutrients to the seed. The plant provides nutrients for the fungus. It's a give and take relationship. When living things depend on each other for survival, we call that **symbiosis**, or a symbiotic relationship. Orchids form this relationship with fungi. When a plant depends on a fungi for its nutrients, we call that a mycorrhizal relationship. That's because *mycor* is Latin for fungus and *rhiz* means roots. The fungi provide nutrients that are taken in by the roots of the orchid. You'll learn all about fungi, mostly mushrooms, in the mycology lesson of this book.

Orchid anatomy is a bit different from other flowers. Examine the orchid diagram on the next page. An orchid usually has an outer fan of three sepals, an inner fan of three petals, and a single large column that has the male stamens attached to the female carpel in the very center. The top fan part that looks like a petal is called the dorsal sepal. Dorsal is a word that relates to the upper part of the back of a plant or animal. Fish have a dorsal fin on their back. If you see a shark, you'll notice its dorsal fin gliding about the surface of the water. That's how you can remember that the dorsal sepal is the top petal-like

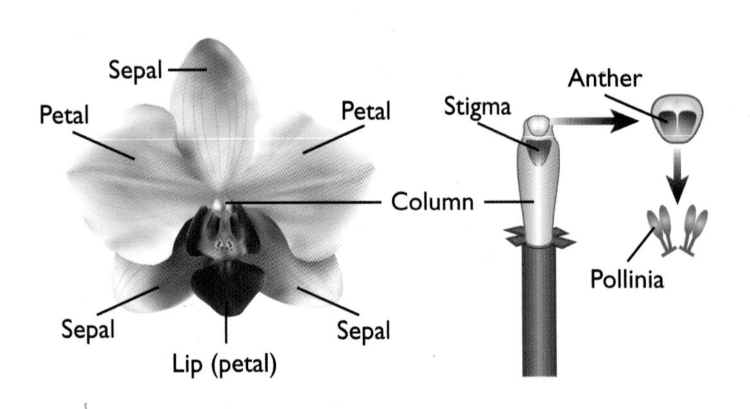

part of the orchid. The outside sepals are called lateral sepals. Lateral means side. So these are the sepals you see on either side of the orchid. You may notice that, unlike many other flowers, orchids are symmetrical. That means both sides of the orchid look exactly alike. We call that **bilateral symmetry**. Living things that have bilateral symmetry can be divided in half and each half looks like the other half vertically. Humans have bilateral symmetry, too.

The very bottom petal-like fan on the orchid is called the lip. Orchid lips come in many different shapes, sizes, and color designs.

In between all these sepals are the pollen-making parts of the flower. Not every orchid has the exact same structures in the center. But they all have stamens and carpels.

Orchids are not just unusually interesting flowers. They also serve many purposes. People use them for making perfume, as well as some kinds of spices. In fact, an important ingredient we use in many baked goods comes from an orchid. Vanilla! Vanilla comes from a species of orchid called *Vanilla planifolia.*

The next time you go to a big hardware store or plant center, look for the orchid section and study the varieties being sold. Some botanical gardens have an indoor section with many kinds of orchids. Schedule a visit to go there. I think you'll enjoy seeing them in real life!

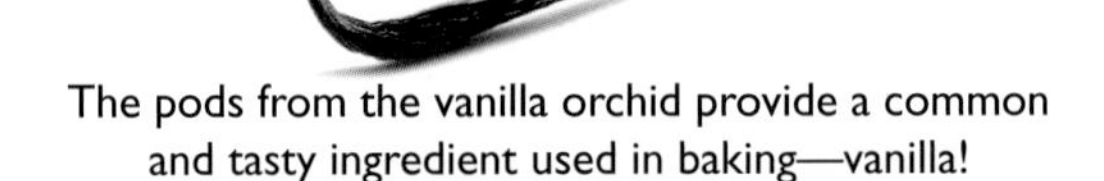

The pods from the vanilla orchid provide a common and tasty ingredient used in baking—vanilla!

ACTIVITY 3.5

LABEL AN ORCHID

In your Botany Notebooking Journal, draw a picture of an orchid and label its parts according to the diagram above. Write down the interesting things you learned about orchids.

FLESH-EATING FLOWERS

Do you know what a carnivore is? It's an animal that eats meat. The root word *carn* is a Latin word meaning meat. Cats and dogs are carnivorous. Can you think of any other carnivorous animal? Animals that eat grass are herbivores. Most carnivores eat herbivores.

The last few flower families we're going to examine are actually carnivorous flowers. Yep. Flowers that consume meat. I know I told you that plants are producers, not consumers. But these odd little flowers are the exception to that rule. They are both producers and consumers. Now don't worry; none of these flesh-eating flowers could or would ever eat you. Most of them are tiny, only a few inches high, but a few are big enough to ingest a small frog or a hummingbird. Though these plants act just like any other angiosperm in every other way, they are different because they're able to absorb certain nutrients from the animals that get lured into their traps.

Carnivorous plants get the nutrients they need by capturing and digesting insects and small animals.

Carnivorous plants actually have special mechanisms inside them that cause visiting insects, and sometimes small amphibians or birds, to get stuck inside them. Sometimes it's a trap, and sometimes it's sticky goo that the animal gets stuck in. Once these unfortunate visitors have been caught, plants digest them to get an important nutrient called nitrogen. All plants need nitrogen, and they are usually able to get plenty of it from the soil. Carnivorous plants, however, are able to live in poor soil or swamp areas where nitrogen is not readily available. God designed these carnivorous creations to make up for the lack of nitrogen in the soil by getting it from small animals in the area. That was a very imaginative idea, wasn't it? There are several families of carnivorous plants. Let's look at some common ones.

THE VENUS FLYTRAP

The **Venus flytrap** is probably the most famous of carnivorous plants. The first Venus flytrap was discovered in North America on the coast of North and South Carolina, where it grows wild near the Cape Fear River. You can easily buy these plants today to keep in your home, as they are found in plant stores all over the country.

If you look for a Venus flytrap in the store, don't look for anything big. The Venus flytrap is a small plant, with tiny little leaves. Its clam-shaped leaves are lined with bristles on each edge and put out a sweet aroma to attract insects. Inside the leaves are tiny, almost microscopic hairs that respond to touch by snapping the leaves shut. If one hair is touched, such as with a stick, the leaves remain open. But if two hairs are touched (one after the other), the trap is triggered and the leaves close. The fact that the leaves require two hairs to be touched protects the plant from snapping shut when a

Keeping a Venus flytrap in your home is an easy way to observe carnivorous plants in action.

nonliving thing touches it. Only living things move around enough to cause two hairs to be touched one after another. The Venus flytrap consumes ants, flies, moths, beetles, grasshoppers, and worms. People have tried to give their Venus flytrap pieces of hamburger, but that's not good for the plant. The plant is designed to use entire animals, not just the meat from the animal.

After it traps the insect with its tiny, sweet-smelling leaves, the Venus flytrap takes about three days to digest the entire insect. After that, the leaves open up again, ready for a new creature to happen upon them. Now it's important to understand that the Venus flytrap does not eat insects for food. Like all plants, it makes its own food to eat. The Venus flytrap, like all carnivorous plants, uses the animals it digests as a source of nitrogen, which you can think of as a vitamin for the plant. Do you take vitamins? They are not your food, but they help to keep you healthy, right? Well, plants need vitamins, too, and the Venus flytrap gets some of its vitamins from the insects it digests.

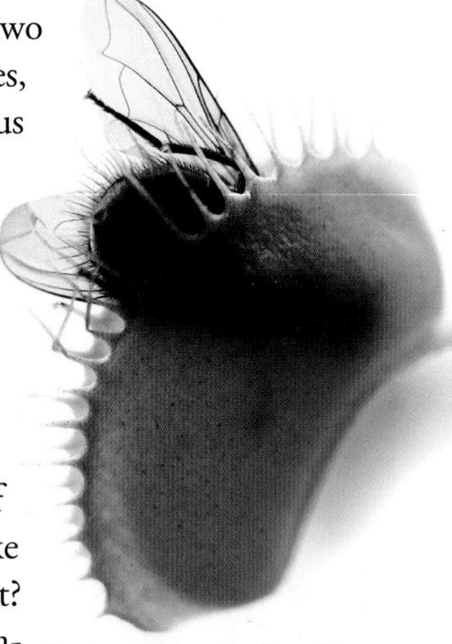

In three days this fly will be completely digested, providing the plant with a very important nutrient called nitrogen.

THE BLADDERWORT FAMILY

A **bladderwort** is a carnivorous plant that looks perfectly normal from above. It grows small, beautiful flowers on its long, bare stems. If you look down below (where roots belong), however, you'll find multiple tiny traps that hang in the water. These traps are actually sacs that act like vacuums, sucking in tiny water critters. As you probably guessed, these plants grow in swampy, wet areas. Their sacs are called bladders. That's why we call the plant a bladderwort. Their bladder-like traps are tiny, less than an inch small. They're sometimes as small as a pinhead.

Within the bladder, a trapdoor is held closed by a thin film of a glue-like substance that blocks the entrance. Special trigger hairs near the lower edge of the trapdoor cause it to open when a creature hits against them. Immediately, in less than a second, the creature is sucked inside the bladder like a vacuum cleaner sucks in dirt from the carpet. After that, little glands inside the bladder release chemicals that digest all the soft parts of the creature. Bladderworts trap and eat water fleas, worms, and small insects. The larger bladderworts can even catch tiny fish!

Underneath these beautiful flowers are traps that suck in tiny water creatures that the plants will consume.

Bladderworts live in nearly every country of the world. They prefer areas that lack enough of the vitamins that plants need. Since bladderworts get their vitamins from the creatures they catch, they can live in such places. However, noncarnivorous plants find it much harder to live in these places.

THE PITCHER PLANT FAMILY

Perhaps the most frightening member of the carnivorous plant group is the **pitcher plant**. This plant is frightening because it's known to consume larger animals. It's a plant that develops vaselike

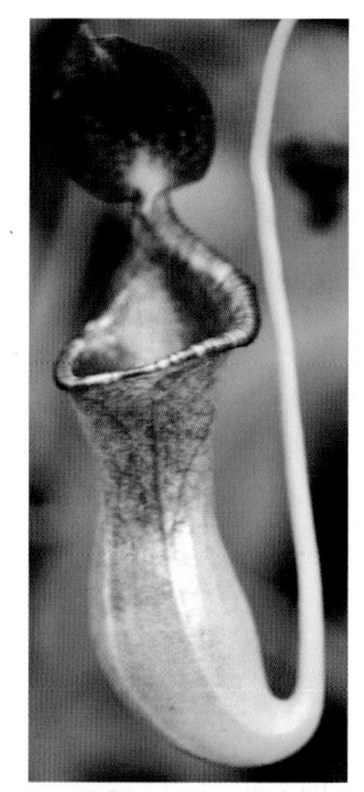

Pitcher plants are designed like vases to capture larger animals.

Pitcher plants come in various shapes, sizes, and colors.

tubes that grow straight up from its grasslike stems. There are many different species of pitcher plants. One species has a hoodlike cover over the vase. In fact, a common American species looks very much like a flower vase.

Inside the vase is a cup of sweet-smelling nectar that attracts bugs and small animals to its lip. When the animal tries to get a little closer for a delicious sip, it slides down the extremely slippery sides of the vase and finds itself trapped inside. Why can't it just climb out? After all, a bug can crawl up almost anything. It's because there are small prickles pointing down into the base of the cup that continually poke the creature when it attempts to climb. In addition, the vase is filled with rainwater. After struggling for several minutes, the creature drowns and within three hours is completely consumed by the pitcher plant. In the end, there's nothing left of the little creature but the exoskeleton (the hard, shell-like body of the insect) or bones. These bones and exoskeletons just stay there at the bottom of the vase.

God designed the pitcher plant in such a way that once an animal enters the plant, it is unable to escape.

Though the pitcher plant was created to consume insects, some birds, frogs, and rodents have been known to fall into the vase and be consumed by the plant. It's very difficult for the plant to consume such large animals, and it often weakens the plant. After all, it's designed to get the nitrogen it needs from insects. It really would rather not have to digest big animals. However, if a big animal falls in the vase, there is no way for the pitcher plant to get rid of it, so the pitcher plant is forced to digest it.

THE SUNDEW FAMILY

Sundews are another amazing creation of God. These spiny, flowering plants grow to be about five inches tall. That's probably as tall as your mom's hand. Sundews are found everywhere in the world except Antarctica and come in many colors. On the ends of each flower stalk are tentacles that ooze wonderful-smelling, sticky goo. This goo oozes down the plant and smells so wonderful that it attracts sugar-loving creatures day and night. Of course, this sweet-smelling goo is actually a trap. It's made of chemicals that not only ensnare insects but also digest them so that the plant can get its vitamins from the insect.

Here's how it works: a fly, butterfly, or other nectar-loving insect detects the sweet smell of the liquid that oozes from the plant. Landing on the colorful tip,

Sundews are beautiful, sweet smelling traps that ensnare unsuspecting insects.

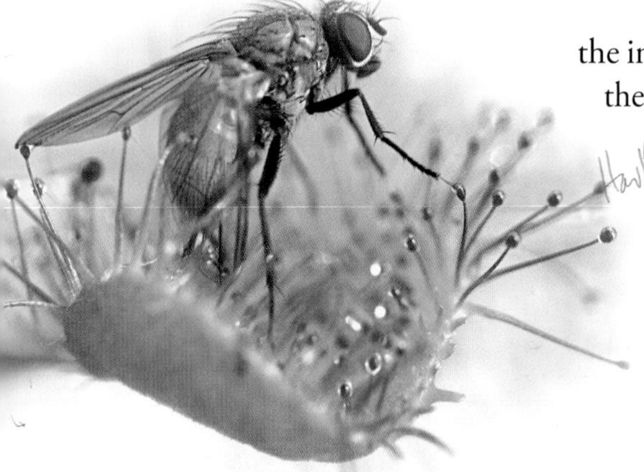

This fly is attracted to the sweet-smelling liquid oozing from the sundew's colorful tips.

the insect immediately gets stuck. The plant has cells that send messages to the plant when an insect arrives. This tells the plant to make more goo in the area where the insect is stuck. The plant then produces a massive amount of goo from the nearby tentacles. The goo completely surrounds the insect. The sundew plant then slowly wraps its tentacles around the insect. That sounds pretty gross, right? Well, at this point the sundew absorbs the nutrients from the insect, consuming the entire insect except the exoskeleton (its outer shell). Once the plant is done with its job, it opens up its tentacles and drops the exoskeleton to the ground. Thank goodness this plant is so small!

WHAT DO YOU REMEMBER?

What is so special about angiosperms? What is the purpose of a flower? What is the job of the sepal? What are all of the sepals together called? Explain what the corolla is. What is the male part of a flower called? What is the female part of the flower called? Where are the flower's ovaries? What do ovules become after they are pollinated? Tell something you learned about flowers in the daisy family, flowers in the orchid family, and carnivorous plants.

ACTIVITY 3.6
DESIGN A FLOWER

After writing down the interesting facts you learned about flowers, think up a design for your very own flower based on what you learned about orchids and carnivorous plants. You could create a new kind of plant called a carnivorous orchid. Think creatively. What would your flower resemble? How big would your plant be? What colors might it have? How would it trap animals? What kinds of animals would it ensnare? Draw a depiction of your creation in your Botany Notebooking Journal and write down its features.

ACTIVITY 3.7
PRESERVE A FRESH FLOWER

You can preserve the color of a leaf or flower by completely surrounding it with borax in a covered container for about two weeks. When you retrieve the flower, it will be the same color it was when you placed it in the borax, but it will be dried out! The color will last a long time.

You will need:
- Fresh flower
- Container with a lid
- Borax

You will do:
1. Pour a layer of borax into your container.
2. Place the flower in the container and completely cover it with the borax.
3. Close the lid to the container.
4. In two weeks, open the container and gently remove your dried, preserved flower!
5. Add it to your nature journal or use it to make a special card to give to someone.

LESSON 4

POLLINATION

digging deeper

Afterward he was traveling from one town and village to another, preaching and telling the good news of the kingdom of God.
Luke 8:1

The way God designed pollen to spread from flower to flower reminds us of the way He wants us to spread the good news that Jesus came to forgive all our wrongdoings and give us a place in heaven with Him. Just as there are so many ways God designed pollen to spread, there are also many ways we can share the good news of Jesus with others.

POLLINATION

Now that you know the flower's purpose is to make seeds, we're going to explore this more deeply. In fact, this is one of the most fascinating aspects of botany. It's called pollination. Here's how it works: For a flower to make a seed, pollen must get from the anther to the stigma. In other words, pollen has to leave the stamen and arrive at the carpel. That's what pollination is, in a nutshell. And that's exactly how nutshells are made! It's super interesting to see how God designed this to happen.

The stamens on this flower are covered with pollen.

How does pollen get from the stamen to the carpel? "That's easy!" you might say. "They are right next to each other." Although that's true, it doesn't help much for most plants. If pollen from a stamen goes to the carpel on the same flower, we call this **self-pollination** because the flower is pollinating itself. But God designed most flowers so that they cannot self-pollinate. For most plants, pollen from one blossom must get to the flower of a nearby plant of the same kind. In other words, the pollen on the stamen from the pear tree in my yard must get to the carpels on the pear tree in my neighbor's yard. But how does this happen?

Actually, there are at least two answers to this question. Wind can carry pollen from one plant to another, but that is not very effective. After all, wind blows in many directions. Pollination from one plant to another will only happen if the wind is blowing just right. Although some plants can get pollinated through the wind, God has designed another, far better, way for pollen to travel between plants. Introducing...animals!

Yes, indeed, animals help plants pollinate one

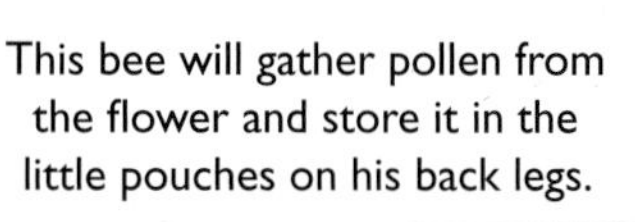

This bee will gather pollen from the flower and store it in the little pouches on his back legs.

While sipping a flower's nectar, butterflies will pick up pollen and carry it to the next flower.

another! Many creatures, such as bees and butterflies, carry pollen from one flower to the next when they're looking for nectar. Remember from the previous lesson that plants make nectar and put it in their flowers. Birds, bees, butterflies, and hundreds of other animals eat this sweet nectar. They must have it in order to survive. And when they are drinking this delectable juice, they get pollen on their bodies. Then, as God designed it, when they move on to the next plant to get more nectar, they transfer pollen to the stigma of that plant's flowers. If it's the right flower, God's plan of pollination happens. It's pretty amazing how God designed the pollen to reach from the stigma down to the ovaries. We'll talk about that in a bit. For now, let's explore the ways God designed animals to help plants reproduce to become new plants.

THINK ABOUT THIS

Dad Our Bible tells us that God is the Creator of everything. What does that mean? Well, look around you. Everything you see, and even the things that you can't see, were made by our God. He made the plants that color our Earth and provide food; He created the birds, bees, bats, and butterflies (all flying creatures). And even more wonderfully, He created the animals to help plants in pollination. You can see how it is part of God's perfect plan to create blooming plants and give them the animals necessary to help them thrive. Do you remember what it's called when two living things are dependent on each other to survive? It's called symbiosis. There's a symbiotic relationship between flowers and many different kinds of flying creatures. After all, it would be much harder for many flowers to complete their special, God-given purpose of making new seeds without animals to help them. Without the help of animals, plants would have a harder time making new seeds, and without a lot of new seeds, the plants would die out.

Without animals like the tiny honey possum to help them pollinate, flowers would not have been able to survive during creation.

Hart Each flower has special colors, shapes, and smells to attract the animals God designed to help the flower spread its pollen. Let's look at some of these animals to see how this works. We'll begin by examining some of the most important pollinators: insects!

BEES

Hart You probably already know that honey bees make honey. Why do they make honey? Because honey is the food they eat during the winter when flowers aren't in bloom. When flowers are available, the bees take nectar into their bodies and some even collect pollen in pollen baskets on their back legs. The nectar turns into honey inside the hive.

Bees make honey to eat during the winter months when flowers aren't in bloom.

During the winter, the bees eat the honey and pollen to survive. Because they use both nectar and pollen, bees are always on the hunt for flowers. They are one of the most important pollinators of many flowers, and because a single bee may visit several thousand flowers in one day, there's a good chance it will visit two of the same kind. When it does, the tiny pollen grains from one flower's stamen find their way to the correct carpel. A baby plant can then be born!

Sometimes the stamen is on a little trigger that flings the anther toward the place where the petal was touched. This throws pollen onto the bee. Other flowers are designed so that the bee must walk by the stigma, rubbing its body against it before it can walk down to the nectaries where the nectar is stored. However, scientists have recently discovered that flowers are electrically charged. Have you ever held a balloon near your hair and it stuck to the balloon? That happened because your hair and the balloon had an electric charge and were attracted to each other. This is also true for flowers and insects. When an insect nears a flower, the pollen actually jumps from the flower to the insect because of this electric charge. God truly thought of everything to make sure flowers get pollinated, didn't He?

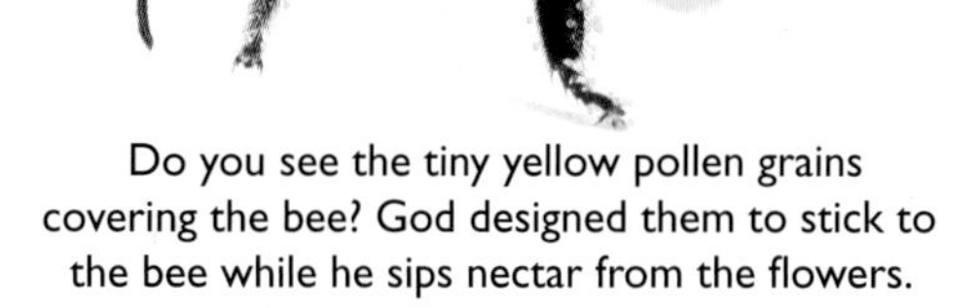

Do you see the tiny yellow pollen grains covering the bee? God designed them to stick to the bee while he sips nectar from the flowers.

Honey isn't just for bees, you know. People have enjoyed bee honey for thousands of years. Some people even keep bees in special boxes in their yard to make honey. These people are called beekeepers. The flowers that are in or near the beekeeper's yard get many little visitors each and every day of the spring, summer, and fall. These fortunate plants are sure to produce many new seeds because the bees help them pollinate each other. Do you see a lot of bees when you go outside? If so, perhaps there's a beekeeper somewhere nearby.

Although bees are really important to food crops, they are not absolutely essential. Many rumors float about that bees are responsible for most of the food we eat, and without them, we would have very little food and eventually all starve. That's an enormous exaggeration and very false. Bees help greatly with food crops. However, the vast majority of crops aren't dependent on bees to pollinate them. Bees can improve the amount of food a crop makes, but they're not the most important factor for most food crops that we depend on for survival. So don't be fooled by those who say the world would be in danger without bees.

Beekeepers use bees and special boxes to make honey to sell locally.

LANDING PAD

Harlan Did you know that when helicopters land, they must have a large flat surface on which to land? This surface is called a **landing pad**. Because bees need a place to stand when they stop to get nectar, God designed some flowers with a good landing pad. You can be sure that if the flower has one, it's likely designed to attract bees.

After a bee lands on a flower, he usually walks toward the center to get the pollen. Once at the center, the bee sticks his head deep inside to gather the nectar and uses his special legs to collect and store the pollen. Of course the flower must be large enough for the bee's head to fit inside.

This flower's landing pad has enough room for both bees.

This butterfly will uncurl its proboscis to sip the flower's nectar.

Hark Like bees, butterflies must have a landing pad, but because butterfly legs are thin, they don't need a large pad. Also, the flower doesn't need to be large enough for the head of the butterfly because the butterfly collects nectar with a long, curly "straw" called a **proboscis** (pro bah' skus). The butterfly lands on the flower and uncurls its long proboscis, which delves deep into the flower and sucks up the nectar.

When you see a flower that is teeny tiny, with a long, thin neck below the opening, you can be certain it's meant for butterflies to pollinate.

SMELL

Harleigh Insects are especially attracted to flowers with a certain smell. Isn't it interesting that God gave insects the ability to smell? Bees prefer nice-smelling flowers while some insects, like flies, are attracted to flowers that smell like rotten meat, such as the corpse flower. Do you remember the bee orchid from the last lesson? It smells like a female bee. So, as you can tell, smell is an important factor in attracting pollinators.

Have you ever noticed there are certain flowers that have no scent during the day but have a sweet smell at night? In the heat of the day, you can't smell these flowers even if you stick your nose inside them. As the day cools and evening approaches, however, they begin to emit a sweet, strong smell. If you take an early morning walk before the day warms up, you can still smell them as you pass by.

Harlan Usually, these special night-aroma plants are very light in color, making them easier to see when it's dark outside. Light colors are easier to see at night than dark colors. That is why you should wear white clothes on evening walks so that car and truck drivers can more easily see you.

Night pollinators don't need a landing pad to collect nectar.

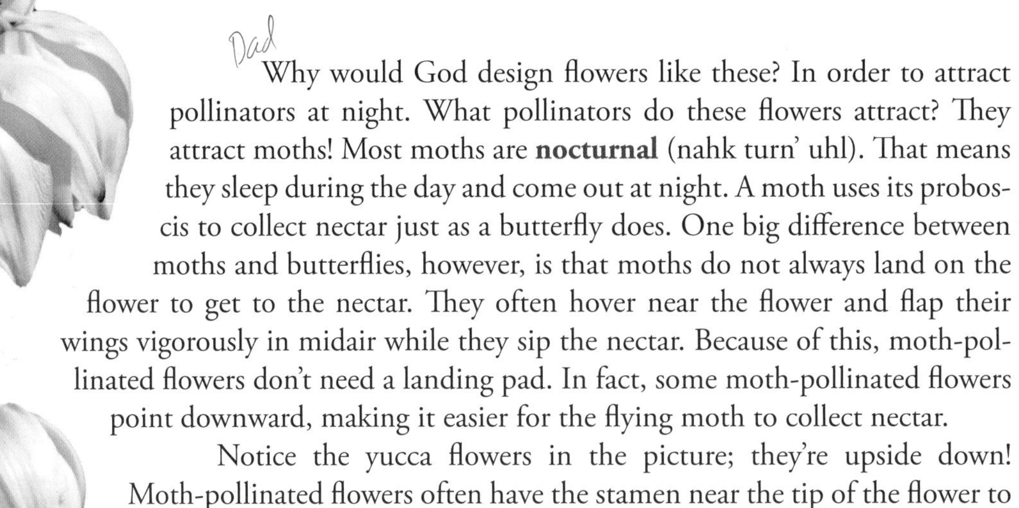

Why would God design flowers like these? In order to attract pollinators at night. What pollinators do these flowers attract? They attract moths! Most moths are **nocturnal** (nahk turn' uhl). That means they sleep during the day and come out at night. A moth uses its proboscis to collect nectar just as a butterfly does. One big difference between moths and butterflies, however, is that moths do not always land on the flower to get to the nectar. They often hover near the flower and flap their wings vigorously in midair while they sip the nectar. Because of this, moth-pollinated flowers don't need a landing pad. In fact, some moth-pollinated flowers point downward, making it easier for the flying moth to collect nectar.

Notice the yucca flowers in the picture; they're upside down! Moth-pollinated flowers often have the stamen near the tip of the flower to make certain a hovering moth gets pollen on it. They also have carpels close to the tip to make it easier for the moth.

Yucca flowers hang upside down, making it easier for hovering moths to pollinate.

COLOR

There's one more thing that attracts pollinators: color! Butterflies are drawn to all colors of flowers. Bees love flowers that are white, yellow, orange, blue, pink, and all colors in between. If the flower is all red, however, bees don't bother much with it unless it smells really nice. Why is that? Because bees can't see the color red! I'll try to explain why.

Did you know that the rainbow contains all the colors we can see? The set of colors in the rainbow (red, orange, yellow, green, blue, indigo, and violet) is called the **visible spectrum** (spek' trum). Visible means it can be seen. Those colors, then, are the colors of light we can see. Did you know there is a lot of light that we can't see? That's right. You and I can only see some of the light that comes from the sun. There are many kinds of light produced by the sun that we just cannot see. Two of these kinds of light are **infrared** (in' fruh red) **light** and **ultraviolet** (uhl' truh vie' uh let) **light**. Even though we can't see them, they are certainly there!

Look at the drawing here. Do you see the rays coming from the sun? Notice that some of those rays form the visible spectrum. Remember, that's the light we can see. However, notice that on both sides of the visible spectrum, there is infrared light and ultraviolet light. This nonvisible light is not seen by human eyes, but it's there. In the drawing, the infrared and ultraviolet light are given colors, but they're not really colored. If we can't see them, how do we know they are there? Special equipment allows us to detect the presence of this nonvisible light.

ULTRAVIOLET
VISIBLE LIGHT
INFRARED

Bees have a different visible spectrum than we do. They cannot see any of the colors at the red end of our visible spectrum. However, bees can see colors very well at the violet end of our visible spectrum. They can also see some of the light that's not visible to you and me. Bees can actually see ultraviolet light! Isn't it amazing God made creatures that can see light we cannot see? If you wonder why He did that, read on!

NECTAR GUIDES

Flowers have special patterns on their petals that tell the bees and butterflies exactly where their nectar is stored. These special designs point the insect toward the center of the flower. We call these patterns **nectar guides**. Have you ever seen a dartboard with a bull's-eye in the middle? That's what a flower's nectar guide seems like to the insect. The bull's-eye is the center of the flower where the nectar is hidden.

Certain flowers have very obvious nectar guides. Pansies have a clear pattern that points the bee directly to where it can find the nectar. Speaking of pointing out where the nectar is, look at the orchid pictured here. It actually seems to have a little arrow pointing to the center of the flower. The arrow tells the bee to go in that direction to find nectar.

This orchid has a clear nectar guide, showing visitors exactly where its nectar is stored.

Although we can see the nectar guides on these flowers, God has designed some nectar guides that only bees and certain other insects can see. This is because many nectar guides are visible only to creatures that can see ultraviolet light.

Now you know why God created bees with the ability to see ultraviolet light! When a bee sees the nectar guides with its ultraviolet vision, it knows where to go to get the nectar.

I want you to get an idea of what these ultraviolet nectar guides look like. Look at the two pictures below. They are photos of the same flowers. The photo on the left, however, was taken with visible light. The photo on the right was taken with ultraviolet light. This is how a bee sees those same flowers. Notice what the bee sees. It sees nectar patterns that we cannot see. This allows the bee to find the nectar easily. You and I might have a harder time finding it. That's okay, though. We don't need the nectar to survive.

OTHER INSECTS

Bees, butterflies, and moths are not the only insects that help with pollination. Wasps, ants, many kinds of flies, and some beetles also act as pollinators. Even lizards can be pollinators!

Would you believe that one of our favorite foods, chocolate, comes from the bean of the cacao tree that is dependent on a tiny little fly called the midge in order to survive? And what would chocolate be without sugar? Sugar makes everything taste sweeter. Well, guess what? A kind of fly called a thrip is one of the main pollinators of the sugarcane plant. So if you have a sweet tooth, thank God for flies!

Lizards are one of the unique pollinators God created!

Isn't it wonderful to know that God created these insects to help increase the beautiful flowers and important food we love to eat? We can thank God that He cares so much about us and also the world He created.

Take some time to tell someone what you learned about insect pollination.

ACTIVITY 4.1

EXPLORE FLOWER POLLINATION

You will need:

- 2 flowers that are still on their plants. They should be on separate plants of the same type.
- Cotton swab (like a Q-tip®)

You will do:

1. Rub the cotton swab in the anthers of one flower. Do you remember what the anthers are? They are on top of the stamens, which are the male parts of the flower.
2. Once you get pollen on your cotton swab, rub the swab on the stigma of the other flower. Your goal is to get some of the pollen from the cotton swab onto the stigma. Do you see how easy it is to pollinate flowers?
3. Watch carefully over the next few weeks to see if your flower loses its petals and the ovary begins to swell. If you succeeded in getting pollen from one flower to the stigma of another flower the ovary will swell into a seed container. As it matures, the seeds inside will mature.
4. Write down what you did in your Botany Notebooking Journal and record the results of your pollination experiment!

BIRD POLLINATORS

Not only do insects help flowers pollinate; some birds do as well. In North America, the hummingbird is an important pollinator of flowers as it flies from blossom to blossom, feeding on the sweet nectar inside. In Hawaii, the honeycreeper is a pollinator. In Australia, the honeyeater pollinates flowers. In some tropical areas, sunbirds and some species of parrots also serve as pollinators.

God didn't give these birds a good sense of smell. So a wonderful-smelling flower isn't what attracts the hummingbird. Instead, they're looking for a certain color as they fly about seeking nectar. Guess what color the hummingbird loves most? Red!

This rainbow parrot is an important pollinator for flowers in Australia.

Although a bee cannot see red, birds can see red very well. When you see a big red flower, you'll know God might have made it to feed hummingbirds, and if it smells nice, butterflies and bees might be invited guests as well.

Like the moth, the hummingbird drinks nectar as it hovers over the flower. Its wings flap so quickly as it hovers that you and I see them as one big blur! A hummingbird might shove its entire head down into the flower to sip the nectar. When it pulls its head out, guess what color its head might be? All covered with pollen, its head turns yellow like a canary's head! Sometimes, the pollen actually gets stuck on the hummingbird's belly. Many times, pollen is simply rubbed onto the hummingbird's little pointy beak. Either way, the hummingbird zooms off to visit the next flower and pollination happens. This is just another one of the delightful ways God has designed flowers to get pollinated while giving life-sustaining nectar to the animals.

God made bright red flowers to attract many pollinators, including hummingbirds!

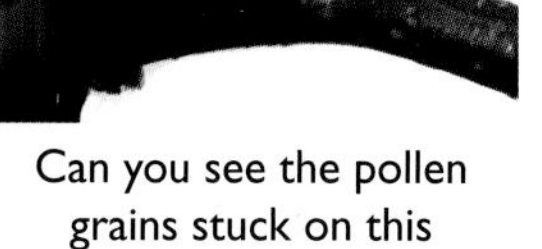

Can you see the pollen grains stuck on this hummingbird's beak?

If you place a hummingbird feeder outside your window, you're likely to see many of these amazing creatures on their search for a sweet drink. It might be fun to plant some favorite hummingbird flowers near the feeder. You can find the kinds of plants that attract hummingbirds on the course website I told you about in the beginning of this book.

ACTIVITY 4.2

BUILD A HUMMINGBIRD FEEDER

You will need:

- Small mason jar with lid
- Red spray paint
- Adult with a drill or hammer and nail (to punch holes in the lid)
- 1 yard of wire (cut in three equal pieces)
- 1 small piece of wire (for fastening other wires)
- Water and sugar

You will do:

1. Have an adult drill or hammer three small holes in the jar's lid near the edges.
2. Spray the lid red and let it dry.
3. Use the pieces of wire to make the hanger: wrap one piece of wire in a circle tightly around the lid. Secure the other two wires to the wire circling the lid, across from one another. Fasten the two wires together at the top with the small piece of wire.
4. Fill the jar with hummingbird food (recipe on next page).

Continued on next page

5. Screw the lid on the jar.
6. Hang the feeder outside on a bird feeder hook or tree.

The recipe for the hummingbird food is four parts water to one part sugar. To make the solution, heat up four cups of water and one cup of sugar. Stir it well until the sugar has dissolved. You can add red food coloring to increase the attraction of birds if you'd like. If you choose to do this, be sure to purchase an organic brand of food coloring that is safe for birds to consume. Once the solution cools, pour it into your feeder. You might want to hang more than one feeder because hummingbirds are very possessive and won't allow other hummingbirds to feed on a feeder once they have claimed it.

Pollen is sticking to this bat's fur as it feeds on the flower's nectar.

MAMMAL POLLINATORS

Like most moths, bats are nocturnal. That means they come out at night after resting all day. Some bat species pollinate flowers. The flowers they visit are aromatic at night. That means they are really smelly when the sun goes down, much like the flowers pollinated by moths. However, bats are not looking for a sweet smell; they prefer musty or sour odors.

Bat-pollinated flowers tend to be pale, and they typically have a strong odor at night. These plants must be strong and sturdy to support the little mammal as it climbs on the branch of the flower bush. They must also have very few leaves so they don't get in the way of the bat as he seeks the flower. A cactus plant is a great flowering plant for a bat. It has sturdy limbs, no leaves, and big flowers. Of course, the bat has to keep clear of a cactus's prickles, so they don't get stuck in its skin.

Do you like bananas? If so, you can thank bats because they are a main source of pollination for the banana tree.

Believe it or not, the black and white ruffed lemur is one of the largest pollinators in the world. It sticks its long nose inside many different kinds of flowers and transfers the pollen that's stuck to its fur.

In addition to bats and lemurs, there are certain varieties of rodents that pollinate flowers as well. You can see that animals are vital to most flowers and their ability to make new plants. However, there are other methods God designed for pollination. We'll study those next.

As this grey-headed flying fox eats flower nectar, its face is getting covered in pollen.

Explain in your own words what you remember about how birds and mammals pollinate.

ACTIVITY 4.3

ILLUSTRATE ANIMAL POLLINATORS

In your Botany Notebooking Journal are pages for each type of animal pollinator we discussed (bees, moths, hummingbirds, and bats). On each page, draw the pollinator and record the interesting things you learned about it.

WIND POLLINATION

Have you ever noticed that during a certain time of year, you can find a fine, yellow dust on everything outside? What do you think that dust is? It's pollen, of course! The reason it's all over the place is because God designed some plants to get pollinated by the wind.

How are plants pollinated by the wind? Well consider trees. Some trees develop little catkin flowers. Do you see the catkins in the picture on the right? Do they look familiar to you? Most people don't realize that a catkin is a flower. It's a male flower because it has only stamens. It doesn't have carpels. The abundance of catkin flowers a tree produces is often amazing. Because they blend in quite well with the tree, we often don't realize how many there are. Have you noticed the hundreds of little catkins that lie upon the ground in early summer?

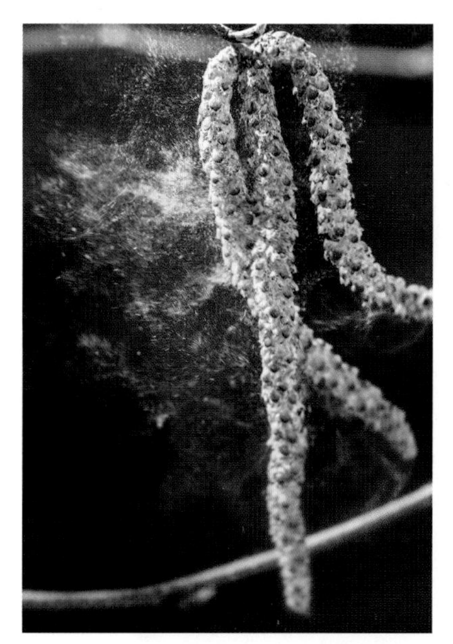

God designed catkins to be pollinated by the wind as the breeze blows pollen off the flowers and onto other trees.

After the catkins form and pollen covers their tiny stamens, the pollen blows off the catkins with every passing breeze. It goes gliding through the air, much of it eventually landing on the ground. Some of the pollen, however, may just blow onto another tree of the same kind nearby. That tree might have female flowers with little sticky carpels (but no stamens) ready to receive the pollen that blows by.

The catkin produces much more pollen than a regular flower because it needs to cover the air with pollen in hopes it will land on a nearby tree. After all, since bees tend to look for only one type of flower at a time, a bee-pollinated flower does not need to produce a lot of pollen. The bee efficiently carries the pollen to another flower of the same type. Since wind pollination is not as efficient, wind-pollinated plants must produce a lot more pollen. God designed the catkin stamens to produce an overabundance of pollen to ensure there will be another generation of trees born from the seeds.

Flowers that are wind pollinated don't have to be pretty, and they usually aren't. Grass flowers and the flowers of many trees are not lovely to behold. Often they're rather greenish or brown in color, the same as the rest of the plant. Most people don't even realize that the little tips on the grass in their yard are flowers. They can sometimes look like little green wheat grains or helicopter blades sitting on your grass.

WHY MOST FLOWERS DON'T SELF-POLLINATE

As I told you before, most flowers cannot self-pollinate. The pollen from one plant must be transferred to the carpel of another plant in order for pollination to actually occur. How did God design flowers to keep from self-pollinating since the stamen and carpel are right next to one another? Not surprisingly, He did it in several different ways. In some plants, God designed the male and female flower parts to develop, or mature, at different times. That way, the carpel on a plant will not be ready to accept pollen when the stamens on the same plant release their pollen. On another plant of the same type, the carpels will be ready to accept pollen before its stamens are ready to release pollen. That way, when the first plant releases pollen, it can only be accepted by the carpels on the other plant.

This water lily cannot pollinate itself because its carpels and stamens mature at different times.

God keeps other plants from self-pollination by giving the plants an allergy to their own pollen. The plant's carpel will happily accept pollen from another plant of the same type, but it cannot accept its own pollen because it's essentially allergic to its own pollen! The tobacco plant is like this. It recognizes its own pollen because of special chemicals in the pollen, and it actually rejects it. However, tobacco plants all produce pollen with slightly different chemicals, so a tobacco plant is not allergic to the pollen from other tobacco plants.

The holly is one plant that produces imperfect flowers.

In some plants, like holly plants, there are stamens on one plant and carpels on another. Because of this, a holly plant that produces pollen cannot accept pollen because it has no carpels. In the same way, a holly plant that has carpels will never produce pollen because it has no stamens. Therefore, it's impossible for a holly plant to pollinate itself. If a plant produces flowers with just stamens or just carpels, the flowers are called **imperfect flowers**. Even though they're named imperfect, they're no less ideal than other flowers. A holly plant's imperfect flowers are perfect for the holly plant.

Here's something important to understand about plants that cannot self-pollinate. Consider roses. Many species of roses will not self-pollinate. That means if I have the only rose bush in the entire city, there will never be seeds made for any more rose bushes. If I want more rose bushes, I'd better plant more than one to begin with. That way, one rose bush can pollinate the other one. The roses will then develop into fruits, called "rose hips," which will contain seeds for more rose bushes.

Rose bushes need other rose bushes nearby in order to be pollinate.

SELF-POLLINATION

Self-pollinating crops are God's provision for people all over the world.

Though most plants cannot self-pollinate, God did design a few varieties of plants to be self-pollinators. That means a single plant *can* pollinate itself! Of course, a plant that self-pollinates can also be pollinated by another plant, so a plant that can self-pollinate does not have to self-pollinate. However, self-pollination is obviously the easiest way to get pollinated. God created only some plants to self-pollinate, and wouldn't you know it, many of these plants produce the foods we eat!

Yes, indeed! God designed self-pollination to help out mankind. It's the method of pollination for many food crops. Wheat, barley, rice, and oat plants can all self-pollinate. The plants of beans, peas, soybeans, peanuts, eggplant, lettuce, peppers, and tomatoes can as well. These are all foods that people around the world depend on to survive!

That means if I only had one wheat seed, I could still grow an entire crop of wheat after a few years. It's the same with the other self-pollinators. God sure takes care of people, doesn't He? The Bible tells us that man is the most important of all of God's creation, for we are the only created thing made in the image of God. He loves us all deeply and desires for us to survive the difficult times here on Earth. He has provided for us in many ways, including making our food easier to grow!

God provided for our nutrition when He created lettuce to self-pollinate.

Wheat is an important self-pollinating food crop.

THE POLLINATED FLOWER

Once a flower has been pollinated, the petals, now finished with their work of attracting guests to their home, dry up and fall off. The flower can then spend all its energy manufacturing the seeds within. Another important reason God made the flower petals die after pollination is so all the other flowers on the bush get an equal chance to be pollinated. If flowers still looked lovely after they were pollinated, birds, bees, and butterflies may spend their visit on flowers that don't need pollination. Fewer seeds would be made if this happened. God really thought of everything, didn't He?

Flowers drop their petals after they have been pollinated.

These ripe ovaries are making seeds to encase in the fruit.

As the petals fall off, the carpel's ovary begins to ripen, making seeds and encasing those seeds in a fruit. The fruit protects the seeds, but it also helps the seeds get away from the parent tree. After all, if the seeds from a tree were to just fall on the ground next to the tree, it would be hard for them to grow. Their parent tree is using the sunlight, soil, and water in that area. It would be easier for the new tree to grow if it were far from its parent. How does the fruit accomplish this? You'll learn all about that in the next lesson.

WHAT DO YOU REMEMBER?

What are the things that attract insects to flowers? What color is the hummingbird most attracted to? Name some mammals that pollinate plants. Explain wind pollination and self-pollination. Can you also explain why a flower petal dries up and falls off after it has been pollinated?

ACTIVITY 4.4

ILLUSTRATE WHAT YOU LEARNED

In your Botany Notebooking Journal, make illustrations of some of the most interesting things you learned today. Title your pages and, if you can, write down what you want to remember about each thing you learned.

ACTIVITY 4.5

CREATE A COMIC STRIP

Have you ever seen a comic strip? It's a story told inside little boxes. If you have a newspaper, you can see what a comic strip looks like. Today you're going to make a comic strip about a flower waiting to be pollinated. Inside your Botany Notebooking Journal, there is a page with boxes for you to create your comic. Each box should contain an illustration and some words explaining the illustration. For example, your first box might contain a drawing of a beautiful flower with an explanation saying something like, "Rosie the Rose waited all day for a bee to come and pollinate her."

ACTIVITY 4.6

MAKE A BUTTERFLY GARDEN

To make a butterfly garden, you'll need to grow the plants that meet the needs of the butterfly in each stage of its life: the egg, caterpillar, chrysalis (cocoon), and adult butterfly. Butterflies lay their eggs only on plants that its caterpillar will eat. Each species of caterpillar eats different plants. However, the butterfly doesn't eat while in the sheltered environment of its chrysalis. It usually hangs from a twig and is hidden from view by its coloring.

Each type of butterfly caterpillar must have a particular kind of food, so you must learn which butterflies are common to your area. If you have trouble learning the species common to your area, and you live in the United States or Canada, you can build a garden for these very common species: painted ladies, swallowtails, whites and sulphurs, gossamer-wing butterflies, brush-footed butterflies, and skippers. Butterflies must have sunlight to warm their bodies. If it gets too cool, they are unable to move. Because of this, it's best to plant these flowers in a sunny place.

Once you learn which types of butterflies are common to your region, you'll need to grow two types of plants. The first is the kind of plant that produces nectar. The second is the kind of plant on which the butterflies will lay eggs. These would be plants the caterpillars enjoy eating. Since each species of butterfly tends to lay its eggs on specific plants, you need to make sure you get the right kinds of plants. Your local nursery will be happy to help you, as many people ask them the same questions you'll need answered. I've included some information to help you.

You will need:

- Vermiculite
- Compost
- Peat moss
- Flowering plants that produce the kind of nectar most butterflies enjoy eating (butterfly bush, lantana, zinnia, bee balm, purple coneflower, penta, sage, milkweed or butterfly weed, lilac, sunflower, marjoram)

You will do:

1. Make your soil by mixing equal parts of vermiculite, compost, and peat moss. This combination is rich in nutrients and will make your plants strong and fragrant.
2. Place the flowering plants in your soil and keep them watered.
3. Enjoy watching butterflies come to your garden!

Continued on next page

BUTTERFLY	PLANT	BUTTERFLY	PLANT	BUTTERFLY	PLANT
Painted Lady	Thistle	Greater and Lesser Fritillaries	Violet	California Sister	Life Oak
Tiger Swallowtail	Tulip Treere	Orange-barred Sulphur	Pea Plants, Alfalfa	American Copper	Sheep Sorrel
Spicebush Swallowtail	Sassafras, Spicebush	Cloudless Sulphur	Wild Senna	Eastern Tailed Blue, Orange-Bordered Blue	Legumes
Anise Swallowtail	Parsnips, Fennel, Carrots, Parsley	Question Mark and Zephyr	Elm	Common Blue	Dogwood Flower
Pipevine Swallowtail	Pipevine	Fawn	Birch	Marine Blue	Wisteria, Alfalfa, Legumes
Black Swallowtail	Fennel, Dill, Carrots, Parsley, Parsnips	Southern Dogface	Wild Indigo, Clover	Southern Cloudy Wing, Northern Cloudy Wing	Clover
Common Buckeye	Snapdragon	Great Southern White	Mustard	Sara Orangetip	Wild Muster
Monarch	Milkweed, Butterfly Weed	Julia, Gulf, Fritillary, Zebras	Passion Flower Leaf	Silver-Spotted Skipper	Wisteria
Field Crescent	Aster	Morning Cloak	Elm, Willow, Poplar	Grizzled Skipper, West Coast Lady	Mallows

LESSON 5

FRUITS

digging deeper

Jesus tells us, "I am the vine; you are the branches. The one who remains in me and I in him produces much fruit, because you can do nothing without me."
John 15:5

What does it mean for us to bear fruit? Have you ever heard of the fruit of the Spirit? *"But the fruit of the Spirit is love, joy, peace, forbearance, kindness, goodness, faithfulness, gentleness and self-control."* (Galatians 5:22–23). Jesus wants us to know that we need His Spirit to produce in us these "fruits" that are described in Galatians. To do that, we should talk to Jesus every day, asking Him to forgive us, to be our guide, and to make us more like Him. Like branches on a vine, we must stay attached to the vine to produce fruit.

A FLOWER'S FRUIT

You learned a lot about flowers in the previous two lessons, didn't you? You learned how a flower makes seeds as a result of pollination, and you learned all about God's special pollinators. But wait! If you can believe it, there's more to this amazing process. After the flower gets pollinated, it has one more job to do. It has to grow its seeds inside a fruit!

Here's what happens: After pollination, the flower's ovary begins to swell and swell and swell while the seeds within it mature. As the ovary swells, it becomes a fruit. Sometimes the fruits are flat pods, such as a pea pod. Sometimes the ovary gets large and plump, forming a fruit like a plum or a grape. Other times it grows a furry tuft or a parachute that will help the seeds float in the wind like the dandelion's seeds do. The ovaries also form into little hard shells, such as peanut shells. Regardless of how a plant's ovary develops into a fruit, the fruit is the container for the plant's seed. Do you remember that angiosperm means seed container? We're going to study these special seed containers now!

These plum fruits are containers for the plum tree's seeds.

The dandelion grows furry tufts that help the seeds float in the wind.

FRUITS AND VEGGIES

It doesn't seem like the shells housing sunflower seeds are fruits, but they are!

Here's something you should know. When the ovary forms a container to house the seeds, we call these containers fruits, even though we don't often think of pea pods and nuts as fruits. When someone says fruit, we usually think of things like apples and oranges. However, everything that grows from a pollinated flower and contains seeds is actually, scientifically, considered a fruit. This means that an acorn is a fruit. It means that the shell housing the sunflower seed is a fruit as well! Many things we would never think of as fruits are indeed fruits, botanically speaking. This is a very important point. Sometimes we use words in our everyday language that are not scientifically correct. For example, most people think of tomatoes as vegetables, but they are not vegetables. What do you see when you cut into a tomato? You see seeds! That means the tomato is a container for seeds, which makes it a fruit. Try to remember that everything that serves as a container for the seed is a fruit.

Most people think tomatoes are vegetables, but they're actually fruits because they are containers for seeds.

If many of the things we think of as vegetables are actually fruits, I bet you're wondering what exactly a vegetable is. A vegetable is any edible part of a plant that does not have seeds. For example, lettuce leaves are vegetables because they do not contain seeds. The buds and stems of a broccoli plant are also vegetables because they have no seeds. Many of the vegetables we eat are actually from the roots of plants. Carrots and potatoes, for example, are parts of the root systems of their plants. Once again, because they have no seeds, they're also vegetables.

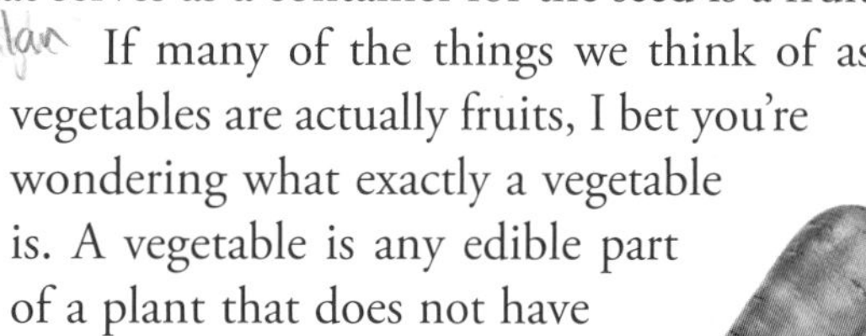

Carrots are vegetables because they have no seeds.

FRUIT KINDS

Now that you know the proper way to distinguish between fruits and vegetables, let's go back to talking about fruits. Remember, a fruit is a vessel that protects seeds as they grow to maturity. In keeping with God's creativity in creation, He, of course, designed many different kinds of fruits. But we can separate most of them into groups. Let's take a look.

God created both fleshy and dry fruits. As you look at the list and pictures that follow, try to think of other fruits you've seen that would fit in the same category as the ones pictured.

Because broccoli is seedless, it's considered a vegetable.

FLESHY FRUITS

Fleshy fruits are fruits that have a fleshy part between the fruit's covering and the seeds. Most of the foods you think of when you hear the word fruits (apples, oranges, grapes, watermelon, etc.) are fleshy fruits. Generally, when you eat one of these fruits, you're eating the fleshy part. The white part of an apple, for example, is the fleshy part of the apple.

We'll learn in a little while how animals are used to spread many different kinds of seeds. However, one thing that's especially interesting about fleshy fruits is the way God designed them to attract animals when the seeds are mature and ready to be released into nature. One way the fruits attract animals is through smell. When the seeds are mature, the fruit becomes ripe. It's the ripe smell that draws the animals and insects. I'll explain why this is important later in this lesson. In the meantime, let's do an experiment to see this in action.

Apples are one of the delicious fleshy fruits God created for us to enjoy.

ACTIVITY 5.1

OBSERVE INSECTS ON A BANANA

In the evening, set two bananas outside your house near one another. Make sure one banana is overripe (blackening) and one is underripe (light yellow with no brown spots). The next morning, check the bananas. Which one attracted the most flies and gnats? How did the insects know the banana was there? It was the smell!

FLESHY CATEGORIES

Fleshy fruits are generally put in the following categories:

An avocado's thin skin and seed inside make it a berry.

Berry

If the entire fruit is fleshy, except for perhaps a thin skin, we call the fruit a berry. A berry might contain one seed or many. Grapes, avocados, and blueberries are examples of berries. They all have a thin skin, but most of the fruit is fleshy. Believe it or not, strawberries are not really berries. This is because the seed or seeds must be inside the berry, but strawberry seeds are on the outside.

Pepo (pee' po)

A pepo is a modified berry. Its skin is hard and thick and is usually called a rind. Pumpkins and watermelons are examples of pepos.

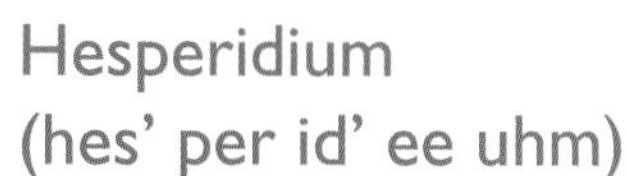

Hesperidium (hes' per id' ee uhm)

A hesperidium is another modified berry. It has a leathery skin that's not as hard as the skin of a pepo. All citrus fruits like oranges and lemons are in this category.

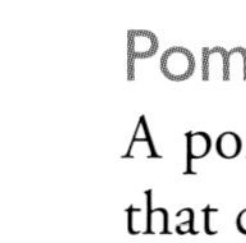

Pome (pohm)

A pome is a fruit that develops a core surrounded by a fleshy tissue that can be eaten. The core is usually not eaten. This is different from a berry, in which the seeds are embedded in the fleshy part, not separated from it by a core. Apples and pears are pomes.

Drupe

A drupe is a fleshy fruit with a hard stone surrounding the seed. We often call this stone the pit of the fruit. Peaches and olives, for example, are drupes. It actually turns out that the almond fruit is a drupe as well. When you eat an almond, however, the fruit has actually been removed. What you eat is the seed, or the pit of the almond fruit.

ACTIVITY 5.2

CATEGORIZE FLESHY FRUITS

In your Botany Notebooking Journal write down all the fleshy fruits you can think of. In addition, write down which fruits belong to which category, then explain to someone what you've learned so far.

ACTIVITY 5.3
SPLIT A SQUASH

In the Bible, Jesus tells a story that says sharing the good news about Him is like planting seeds (Matthew 13:3–8). Take a minute to read in your Bible what Jesus said about planting. Why do you think He told this story? How can it be that if I tell my neighbor about Jesus, I am planting seeds?

You will need:

- Squash or medium-sized pumpkin
- Adult with a sharp knife

You will do:

1. Cut your pumpkin or squash in half.
2. Count how many seeds are inside. Isn't it simply amazing how many seeds this single fruit from one vine can produce?
3. Imagine if one seed could produce a plant that grew 10 more squashes. Try to calculate how many more squash plants each seed could produce.

Dad

How many seeds do you think it took to grow this one fruit? You might be surprised to learn that a single seed produces many fruits. A single pumpkin or squash seed produces a single vine, and a single vine might have several fruits. Each of those fruits will have about as many seeds as the one you cut open. Let's suppose the vine from which your fruit came had a total of 20 fruits on it. How many total seeds, then, did this one seed produce? To answer that question, multiply the number of seeds you counted by 20. Each of the seeds in your fruit could produce roughly the same number of seeds. If you count all the seeds the one seed that was planted to grow this one vine produced, and then you count all the seeds those seeds could produce, how many seeds in all could that one seed make? To answer that question, multiply the number you just got by itself. That's a lot of seeds, isn't it? Of course, the number of seeds produced by that single seed will increase as each seed grows into a new plant that produces more seeds. Can you see how sharing Jesus with one person can be like the one seed that grew your pumpkin?

Hart

THINK ABOUT THIS

Do you think every single seed that a plant produces will grow into a new plant? No, not every one will. The same is true when it comes to sharing Christ with others. Sometimes you'll tell someone about Christ and it won't matter to him. Even so, we should still continue to plant seeds because some of them will certainly grow and produce more seeds that will be planted to grow and produce more and more and more. One day, when you get to heaven, you'll be able to see all the work that was done

Jesus and the Parable of the Sower

through you because you were faithful to plant that one seed that was needed.

How do we plant spiritual seeds? Here are some practical suggestions, but remember God will lead you in specific situations if you ask Him. You can plant seeds by asking people what they think about Jesus and then telling them what Jesus means to you. You can tell them about the prayers He has answered and how He has helped you. When God answers a prayer, tell everyone you see about it. When people tell you they are worried or sad, ask them if you can pray with them. Pray diligently for everyone who needs it. You may not know how many seeds grew from the words you spoke to others, but God knows and is delighted with your planting!

Explain in your own words how sharing Jesus is like planting seeds.

DRY FRUITS

Now that we've thoroughly explored fleshy fruits, let's take a look at dry fruits. Dry fruits are fruits that have no fleshy part between the fruit's outer layer and the seed. Dry fruits are usually put in the following categories:

Grain or caryopsis (ka ree op' sis)

A grain is a very small, one-seeded fruit. The seed is coated in what is called the bran, and the inside of the seed is called the germ. Although we tend to think of grains as just seeds, they're actually fruits because they come from ripened ovaries. Plants that produce grains are the most important food-producing plants in the world. Corn and wheat are examples of grains. Each ear of corn or stalk of wheat contains many grains that are harvested and often ground into a powdery substance for cooking.

Pod or legume

A pod always has a row of seeds attached to the side of the fruit. Pods usually split open along both sides when the seeds are ready. Pea plants, peanut plants, and bean plants all produce pods.

Capsule

Each capsule has several seeds inside. They split open when ready, revealing a group of seeds collected in one hardened ovary. Capsules can split down the sides or around the middle. Some have "windows" that open at the top of the capsule. The seeds fall out of those windows. The fruits of the poppy and primrose plants are capsules.

Follicle

A follicle is a lot like a capsule, but it splits open along only one side when it releases its seed or seeds. Magnolia trees produce fruits that have one seed in each follicle, while milkweed plants produce fruits in which many seeds are in one follicle.

Achene (uh keen')

An achene is a single seed inside a shell; however, the seed is separated from the shell. A grain is also a single seed in a shell, but in a grain, the seed is attached to the shell. In an achene, it is not. Achenes usually form in groups. They can have parachutes attached, like dandelion seeds, or they can have hard shells, like sunflower seeds. Here's something very interesting: The strawberry plant produces achenes. These are the seedlike structures you see on the outside of the strawberry. The seed is inside the achene.

Samara

A samara is a seed inside a winged fruit that floats in the breeze. Samaras that are attached to one another, making a two-winged fruit, are called schizocarps. The samara is also a type of achene since it's only one seed inside a single shell. Maple, ash, and elm trees produce samaras.

Nut

A single seed surrounded by a hard, woody covering is called a nut. Nuts float in water and are often buried by squirrels. They're also harvested in orchards by humans. Acorns and chestnuts are examples of nuts. A peanut is not really a nut because it usually contains more than one seed.

ACTIVITY 5.4

FIND AND ILLUSTRATE FRUITS

Go outside and look for fruits on the plants in nature. Draw them in your nature journal. Write down what kind of fruit you believe it is. This is much easier in the late summer or early fall when plants have gone to seed. But you can still find seeds around in the spring as well.

SEED SCATTERING

We've spent a great deal of time studying the varieties of fruits God created to hold seeds. Eventually, all those seeds we've discussed are released from the fruit and must be spread somewhere to grow new plants. As you probably guessed, just as there are many kinds of seed containers, each seed container comes with its own special way of spreading its seeds to other places to grow.

But why must they spread to grow? Because it's really important for these seeds to find their way to healthy, nutrient-rich soil so they can grow into healthy plants. The problem is the supply of nutrients in each plot of soil is often quite limited and runs out. When a plant grows, it takes these nutrients from the soil and uses them up. If the seeds from a plant simply fell on the very spot where the parent plant grows, the young plant would have to compete with the parent plant for the same nutrients in the soil. And sometimes, one spot of land can only support one healthy plant. The mother plant that produces seeds shouldn't have to struggle with her offspring for the nutrients in the soil. If they were to battle with one another over water and nutrients, the mother plant with her well-developed roots would likely win. As a result, the baby plant would not be able to get enough nutrition and would die or be very unhealthy. On the other hand, if the mother plant was old and weak, the baby plant could grow vigorously and suck up all the nutrients, starving its mother of what she needs, and eventually killing her.

However, the Lord God specially designed His creation so this would not happen. Very few seeds drop straight down from the parent plant onto the same plot of soil. Instead, God built those little seed houses called fruits to move the seeds to other locations. And He gave each fruit a unique way of moving its seed to a new plot of ground. Sometimes the fruit is designed to move the seeds just a small distance away, but sometimes it's designed to move the seeds very, very far away. Let's discover the different ways seeds move from the parent to a new place to grow.

The process of getting the seeds from the parent plant to a new location is called **seed dispersal**. Dispersal means spreading or scattering.

Seeds can be dispersed in five different ways: human, water, wind, animal, or mechanical dispersal. Can you guess how each of these works? Think about each one for a moment and make your best hypothesis (guess). We'll see if you're correct as we read on to explore the wondrous ways God designed seeds to scatter.

HUMAN DISPERSAL

Scientists often ignore human dispersal when discussing the different methods of seed dispersal. Nevertheless, when God created Earth, He planned for many kinds of seeds to be dispersed mainly by humans. He designed these seeds to grow food for people. Farmers raise many plants for food and even raise some plants to give us what we need to make clothing.

Some textbooks say that humans slowly learned how to do this, but that is just not true. The Bible tells us that God put Adam in the Garden of Eden to care for it (Genesis 2:15). So God must have taught Adam how to grow plants and care for them. Adam passed this skill on to his children, who passed it on to their

Growing food is hard but rewarding work.

children, and so on. Before Adam sinned, farming was easier. After he sinned, however, one of the curses resulted in farming becoming very difficult and man needing to work very hard to grow food (Genesis 3:17–19).

As we can see, farming has always been a part of God's great plan for humankind. Of course, all the seeds that are primarily dispersed by humans have other methods of dispersal as well. Even though this is the case, we should remember that God designed farming as a way for us to get much of the food we need to survive. You'll become a farmer of sorts when you create your comestible garden in Lesson 9. I think you'll enjoy growing your own food to eat!

WATER DISPERSAL

Have you ever seen living plants growing in water? The water lily is an example of such a plant. Water lilies have beautiful flowers. When the flowers are pollinated, they create a fruit that floats in the water for a while then drops down to the bottom to take root on the floor of the pond. The seeds of water lilies have been specially created to be dispersed and germinate in water.

There are many other plants that produce seeds that can float, even if the plant doesn't grow in water. These plants use water as their primary method of seed dispersal. Once the seed falls into the water, it will be carried a long distance before it finds a resting place to grow.

This coconut will float about until it eventually finds a place to begin growing.

Coconut palm trees are a great example of a plant God designed to disperse its seeds by water. Have you ever noticed that palm trees often grow near oceans? Ocean currents are powerful and stretch from one continent to another. This is the road the palm tree uses to move its seeds (which we call coconuts) to its new home. When a coconut lands in the ocean, it floats a long time and is sometimes carried thousands of miles away to a new continent. When it arrives at its new home, it will begin growing. The next time you buy a coconut from the store, put it in a sink full of water and watch how it floats.

If you've ever been to Florida and seen mangrove trees, you know they live in water. Their seeds fall from the tree and grow roots as soon as they touch any kind of soil. During low tide, they may fall in soil rather than water and start growing right where they fell. If the water level is high, however, they can be carried far away from where they fell. Mangrove trees are often the beginning of what will one day be a small island. As dirt and debris collect in their roots, little bodies of land are formed.

Mangrove trees live in water and sometimes grow to be small islands.

Most nuts, such as acorns, walnuts, and pecans, are dispersed by both water and animals. Water dispersal can happen if there's a flood or if the plants grow near a river or stream. Many nut trees are found in areas that have a lot of flooding or in areas where they hang over streams or rivers. Although seed dispersal is important to trees, it's not as important for them as it is for other plants. You see, many trees have roots that are able to grow a long way from the tree to find nutritious soil and plenty of water. That's why it's not uncommon for trees, such as oaks and chestnuts, to simply drop hundreds of seeds right around the parent tree. The seeds don't have to compete with roots from the parent tree since the tree's roots are so far below the ground. The main thing the seedling will have to compete for is sunshine. God designed another way for tree nuts to be dispersed. Keep reading to find out how.

ANIMAL DISPERSAL

There are several ways God planned for animals to help in seed dispersal. Have you ever wondered why God created those stickers or burrs that stick to your socks and pants when you walk through grassy fields? Well that's a special method of seed dispersal! Inside each little sticker is the seed that developed from the flower of the plant. The parent plant develops the little stickers or burrs from the flower. These stickers can be tossed off the plant and onto the ground. Or they can stay on the plant until a passing animal gets the burr stuck in its fur or feathers or a person gets it stuck on his socks. The burr is then carried to a new location where the animal or person attempts to get it off. Animals will gnaw, scratch, or peck at a burr for hours to remove it. The creature will then toss the burr on the ground where it can grow a new plant far, far away from the soil that nourishes the parent plant. What a clever way to get the seeds to a new home!

Burrs use animals and people to help disperse their seeds.

Velcro®

Did you know that the person who developed Velcro was actually a scientist named George de Mestral? While on a hunting vacation in Switzerland, de Mestral came home one evening and tried to remove the burrs stuck to his dog's fur. He was shocked at how difficult the burrs were to remove. That night he studied the burrs under a microscope and noted that each burr was covered with hundreds of small hooks acting like grasping hands. De Mestral decided that this God-designed burr could be made to close fabric just like buttons and zippers. By copying the hook pattern he saw on the burrs, de Mestral developed Velcro! Velcro was named after the French words for velvet (velour) and hook (crochet). Velcro is man's imitation of God's handiwork in creation. Did you know that a two-inch square of de Mestral's Velcro is strong enough to hold a 175-pound person hanging on a wall?

Velcro is a material designed after the burrs God created in nature.

ACTIVITY 5.5

EXAMINE BURRS

Burrs are quite amazing. If you are able to find one, examine it carefully under a magnifying glass. You'll see a special network of hooks and latches that are designed to stick onto fur. After you examine the burr, roll it onto a sock then try to remove it. Study it under the magnifying glass to learn why it's not easily removed. You can see now why de Mestral would want to recreate this wonderful design in his lab!

Grass

Many fruits of grasses are equipped with latches that catch on to passing creatures as well. The pointed fruits of spear grass have long, twisted tails (called awns) projecting from their tips. These tails can get stuck in passing animals or can float in the wind to a new location. Once the seed hits the ground, however, moisture causes the awn to straighten and it moves straight down into the soil. After it dries out again, the awn twists back up. This actually screws the seed into the soil where it can germinate. The fruit of spear grass, then, not only helps disperse the seeds, but it also helps plant them!

The awns on this fountain grass can latch on to animals or float in the wind in search of a place to disperse their seeds.

Little Gardeners

Another way God arranged for animals to help in seed dispersal is for the animals to actually plant the seeds themselves. Yes, some animals behave like little gardeners without even knowing it! You see, animals like mice, squirrels, and jaybirds gather fruits and nuts during the spring and summer then store them for food in order to survive the upcoming winter. Often, they store these food sources by burying them under the ground. Most of theses animals are hard workers and store many more fruits and nuts than they need. As a result, some of their fruits and nuts are left behind. Those that are left buried become spring seedlings.

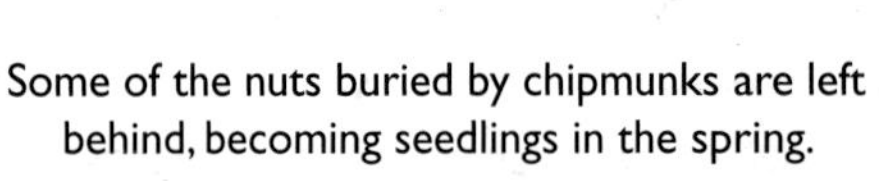

Some of the nuts buried by chipmunks are left behind, becoming seedlings in the spring.

Mistletoe

Birds help with the dispersal of mistletoe in an interesting way. Mistletoe has sticky seeds inside its berries that are attractive to birds. The seeds stick to the birds' beaks when they're eating the berries. The birds then fly away, only to land on another tree where they rub their beaks clean on the tree's bark. The sticky seeds are left on the bark to grow into new mistletoe plants. You see, mistletoe doesn't grow in the ground. It actually grows on trees. Mistletoe is a **parasitic** (pear' uh sih' tik) plant. Parasitic plants steal nutrients from other plants. In the case of mistletoe, it steals nutrients from trees.

Seed Droppings

Have you ever wondered why so many weeds and new plants grow in your yard? It's because God also uses animals to help in seed dispersal by creating lovely, tasty fruits that the animals swallow, seeds and all. They digest the soft fruit, but the seeds go through their bodies unharmed, passing out in their droppings. This is why I have wild strawberries growing in my yard. A bird ate a strawberry then dropped the seeds in my yard when it sat in my trees.

In certain parts of the Earth, especially South America, bats are very important in the dispersal of seeds. Bats that live on fruit can eat up to three times their body weight in a single night. The seeds of the fruits pass through the bats in only 15–20 minutes. The bats then scatter them on the forest floor in their droppings. The short-tailed fruit bat in South America can scatter up to 60,000 seeds in a single night!

The fruit bat's droppings are important to seed dispersal.

Explain to someone all you've learned before you read about the last two methods of dispersal.

WIND DISPERSAL

God has designed some fruits to be dispersed by the wind. What do you think is an important characteristic for a wind-dispersed seed? Well it can't be heavy like a coconut! It must be able to float easily in the breeze and would travel best if it were very light or even perhaps had wings of some sort. Guess what! That's exactly how God designed wind-dispersed seeds! Let's discover some designs for wind dispersal.

God made some seeds to grow a little tuft or parachute on top, like milkweed and dandelion seeds. Other seeds, like orchid seeds, are small and light, almost like dust. Poppy seeds are also very light. They're contained in a little capsule that has windows around the top. On a windy day, the poppy fruit capsule sways from side to side, shaking out the tiny seeds from the windows of the capsule, like a saltshaker. The seeds then float for a short distance to find a new plot of soil where they settle to the ground and grow a new poppy plant. Isn't that amazing? That's why you might

Do you see the tiny seeds in this poppy's capsule?

happen upon a whole field of poppies!

Another kind of fruit God designed to be wind-dispersed is the maple tree fruit. As I mentioned already, the maple tree develops little **schizocarps** (skit' suh karps), which are two-sided winged fruits. They're usually called **samaras** (suh' mah rus) and are light little fruits that fly off the tree with a strong wind. Their wing structures help them stay in the air until they've reached a new plot of ground. Elm and birch trees are also equipped with samaras that have a winglike structure surrounding the seed. When they fall from the trees, they look like little butterflies flittering to the ground.

MECHANICAL DISPERSAL

Have you ever seen a slingshot? It's a device that can fling a rock far away. In 1 Samuel 17:33–51, we read that David used a sling to defeat Goliath. A sling is kind of like the slingshot. Although the design is different, the job it performs is the same. Both a sling and a slingshot can fling a rock much farther and faster than a person can throw a rock. God gave some fruits the natural ability to act like a slingshot and fling their seeds away when they are ripe. This is called mechanical dispersal because the fruit is launched from a little mechanism, or machine.

Pea pods often use mechanical dispersal. When the seeds are ready, the pod dries up. The inside of the pod dries faster than the outside. This causes the pod to twist inwardly, suddenly splitting open with a violent force, rolling into a little spiral. This spiral roll causes the seeds to fly out of the pod in all directions.

Sometimes mechanical dispersal is even more exciting than this. The touch-me-not, which grows wild in the United States, uses mechanical dispersal. The orange flowers that look like little jewels attract hummingbirds and butterflies for pollination. Once the flowers have been pollinated, the seeds form into dry capsules. When the seeds are ripe and ready, the dried fruit is on a God-made trigger. When an animal or human touches the plant, the capsules burst open and spray the seeds everywhere! If it's particularly windy, the plants can hit against one another, bursting the capsules and dispersing their seeds with each gust. If the ground is wet, the seeds might germinate where they land. They might also stick to the creature that set off the trigger, only to be carried off to another location.

The fruits of the violet and the gorse (a European shrub) use mechanical dispersal that's accompanied by a loud noise. The fruit dries into a little capsule. When the seeds are ready, the capsule snaps open with a "POP!" It's been said that sitting in a field of ripe gorse capsules is like sitting in a field with gunfire going off. Each pod can expel up to 8 seeds 9 feet away from the mother plant.

Another fruit that uses mechanical dispersal is the squirting cucum-

These capsules will soon pop, spraying out their seeds and dispersing them in every direction.

ber. Its small, two-inch cucumbers are filled with a slimy juice that contains the seeds. As it ripens, pressure causes the cucumber to burst off its stalk and explosively shoot slimy liquid up to 20 feet away! The seeds spew out with the liquid, and voila! Seeds are sent off to a new plot of land. These would be fun plants to grow, don't you think? If you grow these plants, wait until the cucumber is nice and fat before you try to make it squirt. Gently shake the vine then immediately stand back so you don't get slimed! You should never eat squirting cucumbers; they're not good for you.

There are hundreds of other fruits that use mechanical dispersal. Next time you are outside, especially near the end of the summer and in the fall when fruits are usually ripe, look for capsules and seeds that might snap open at a touch. If you find one, record in your nature journal what it was and what it did. You might even keep a specimen for your journal.

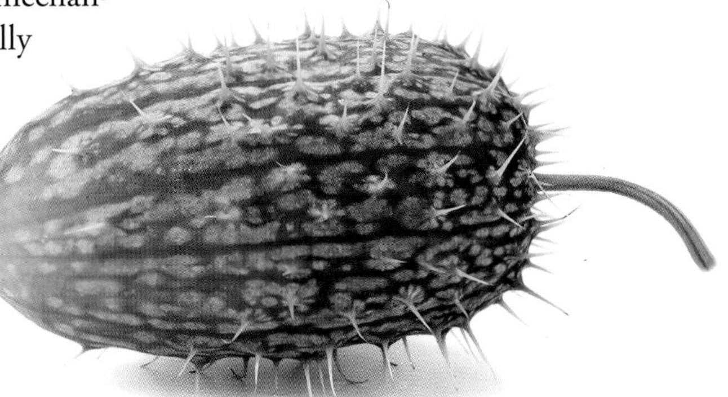

When this squirting cucumber is ripe, it will burst, spraying its seeds out in a slimy liquid!

Now you know why there are so many unwanted plants and weeds growing in your yard and garden. They were dispersed there. Some were dispersed by wind, some by mechanical means, and some by animals.

WHAT DO YOU REMEMBER?

What is the main purpose of fruits? What is the difference between a fruit and a vegetable? Name a few different kinds of fruits that you learned about. Describe what seed dispersal means. Explain the different methods of seed dispersal.

ACTIVITY 5.6

DESCRIBE SEED DISPERSAL

In your Botany Notebooking Journal, you'll find pages titled: Human-Dispersed, Wind-Dispersed, Water-Dispersed, Mechanically Dispersed, and Animal-Dispersed. On each sheet, write down descriptions or illustrate pictures of each of these kinds of seed dispersal.

ACTIVITY 5.7

PRESERVE THE COLOR OF FRUIT

Have you ever noticed that when you don't eat your apple quickly enough, it turns brown? That's

because there's an enzyme in the apple that reacts to the oxygen in the air. But did you know you can use preservatives to retain the color of the apple and prevent it from reacting to the oxygen? Many foods have preservatives in them. Some preservatives are made of natural substances. We're going to do an experiment to see which natural preservative best keeps your apple from turning brown.

You're going to test different substances to see which one preserves the color of fruit. Before beginning this experiment, gather all the supplies and then make a hypothesis (guess) about which chemicals will keep the fruit from turning brown.

You will need:

- Apple cut into 5 slices
- 4–5 cups filled with different substances: lime juice, vinegar, olive oil, water, and/or saltwater. You can also add more cups and try other substances you think may be helpful in preserving the fruit's color.

You will do:

1. Dip an apple slice into each cup to coat it with the substance in the cup.
2. Put each apple slice in front of the cup into which it was dipped.
3. Check back in an hour to see if the apple slices turned brown or if their color was preserved.

Why do some apple slices turn brown while others don't? The food's outer skin protects the interior from rotting. If a fruit or vegetable is injured, the skin breaks, allowing the flesh to become brown and later rot. Acids, such as lemon juice, lime juice, and vinegar, stop the browning because they prevent the oxygen from oxidizing the fruit.

LESSON 6

LEAVES

digging deeper

The person who trusts in the Lord, *whose confidence indeed is the* Lord, *is blessed. He will be like a tree planted by water: it sends its roots out toward a stream, it doesn't fear when heat comes, and its foliage remains green. It will not worry in a year of drought or cease producing fruit.*
Jeremiah 17:7–8

Sometimes we put our trust in things other than God. But if we trust in God, He will make us like a tree planted by the stream, whose leaves are always green. The leaves are green because the tree has a constant supply of water. When we seek God, we are filled with a constant supply of Living Water. For Jesus said, *"The one who believes in me, as the Scripture has said, will have streams of living water flow from deep within him."* John 7:38

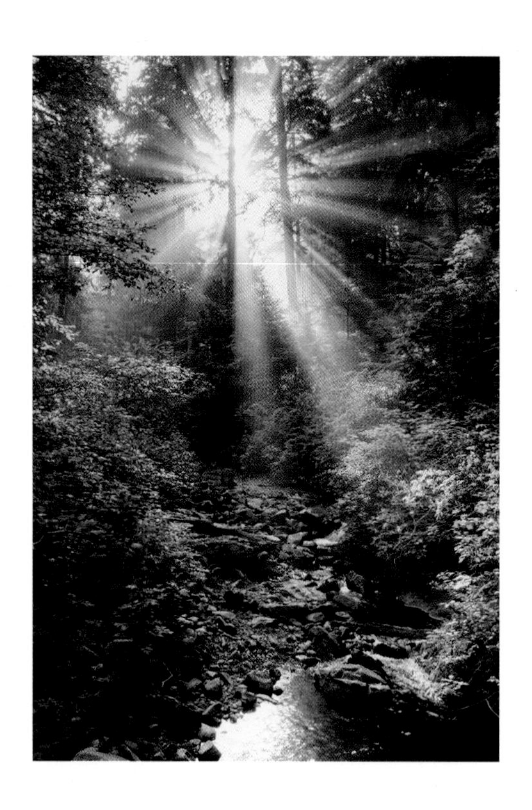

LEAF MOUTHS

Did you know that most plants have a mouth? Believe it or not, they have several mouths. In fact, they have many, many tiny mouths. You probably haven't seen them because they are **microscopic** (my' kroh skop' ik). That means they're too small to see with just your eyes. However, you can see them with the help of a **microscope** (my' kroh skohp). The drawing on the right illustrates what these little mouths look like. Do you know where the mouths are located? Can you guess? I'm sure you know this lesson is about leaves, so you probably guessed they are somewhere on the leaves. You are right! On the bottom side of every leaf are many little mouths, called **stomata** (stoh mah' tah). **Stoma** (stoh' muh) is a Greek word that means mouth. Because of this, one mouth on a leaf is called a stoma, and two or more are called stomata.

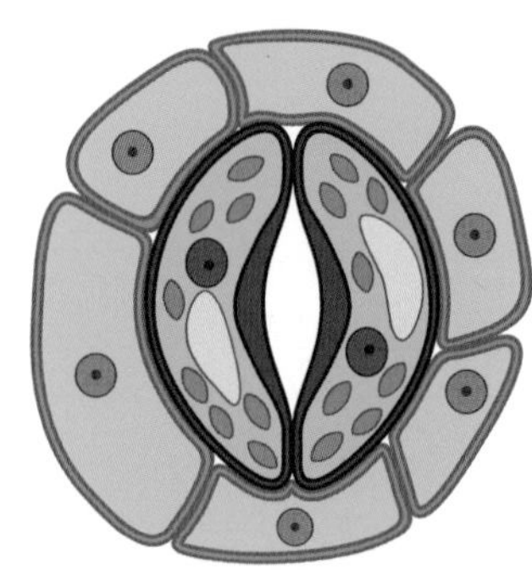

Diagram of an open stoma.

Do you need your mouth to live? Of course you do! Your mouth is as important to your body as the plants' mouths are to them. A leaf's stomata are tiny, but they open and close just like your mouth. These little leaf mouths don't eat food the way our mouths do; they actually eat air. Can you imagine if all you needed to eat was air? Interestingly, that's what plants eat! They take in air to help make the food they need.

As I've mentioned before, most of the living things in God's creation eat (consume) plants or other creatures. As a result, they are called **consumers**. Plants, on the other hand, are quite special because they make (produce) their own food. This is why we call plants **producers**. In this lesson, we're going to learn more about the interesting process by which plants make their own food.

Stomata help the plant make its food by allowing important chemicals in the air to enter the leaf. Without stomata, a plant wouldn't be able to live. Of course, since the stomata are in the plant's leaves, the plant must also have leaves in order to live. When you start collecting leaves for this course, be careful. Take only one leaf from each plant so that the plant will still be able to make food.

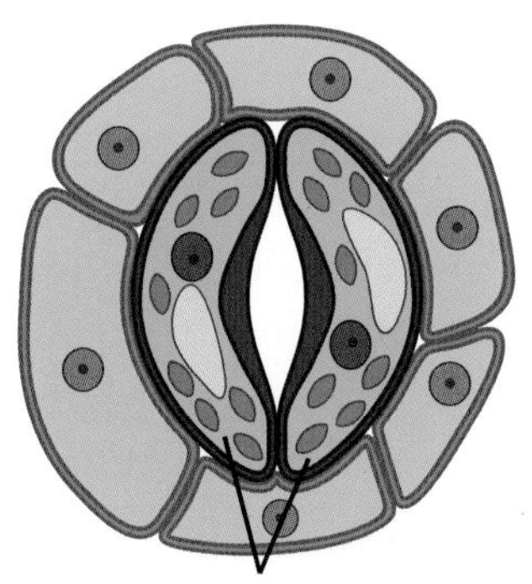

Guard cells

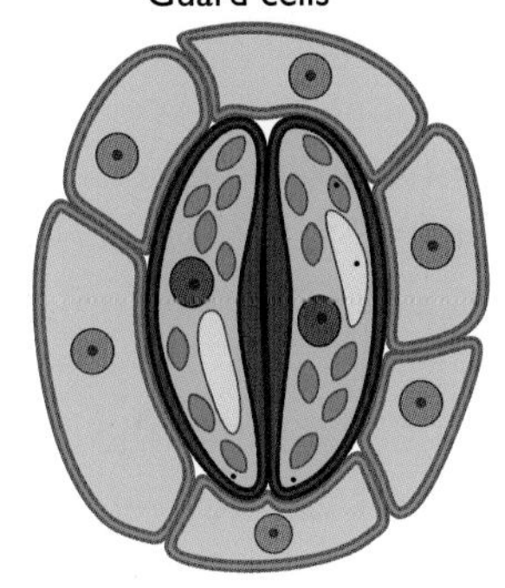

Closed Stoma

The stomata are tiny holes in the leaf. They're surrounded by little structures called **guard cells**. Do you remember what a cell is? The cell is the basic building block of life. All living things are made up of at least one cell. You can think of cells as tiny bricks that are used to build a living creature. If a person put bricks together in a certain arrangement, he could make a house. In the same way, when God puts cells together in a certain arrangement, He makes living things.

Let's talk about guard cells. Each stoma is surrounded by two banana-shaped guard cells, as drawn on the left. When the guard cells expand, they bow outward, away from each other. That opens the stoma. Look at the top drawing. The cells are larger. Notice how there's a large hole in between them. The hole is the stoma. When the guard cells shrink, they end up nestling closer together, which closes the stoma. Look at the bottom drawing. The cells are smaller there. Notice that the hole is nearly gone. It's a lot like how your lips open and close your mouth. Perhaps you could consider your lips guard cells. That makes a lot of sense since God tells us we should guard the words coming out of our mouths.

LORD, set up a guard for my mouth; keep watch at the door of my lips. Psalm 141:3

So here's how the guard cells work: as soon as the sun rises in the morning, the guard cells swell, and the stomata open. This allows a chemical in the air, called **carbon** (kar' bun) **dioxide** (dye ox' eyed), to enter the leaf. The leaf then uses the carbon dioxide to make food for the plant. You'll learn more about how the leaf does this in the next section. When the sun sets, the guard cells shrink. This closes the stomata, and carbon dioxide no longer gets into the leaf. At that point, the leaf takes a break from the hard work of making food for the plant.

You can see why leaves are so important to a plant. Remember this when you're collecting specimens for your nature journal. Don't ever take all the leaves off a plant. When I was a child, my mother had beautiful fern plants lining the entry hall of our home. Every day as I passed by, I would grab one branch of leaves and, beginning at the bottom, pull my fingers up the branch, enjoying the sensation of every leaf popping off. I didn't realize how important the leaves were to the survival of the plant or that I was

Every leaf on this fern is important to the plant's survival.

Tropical rainforests supply 20 to 30 percent of the world's oxygen.

making the plant weaker and weaker as I removed its tiny little mouths each day. Those leaves were vital to the plant's survival.

Believe it or not, the plant's leaves are not only important for the plant's survival; they are also important for *your* survival! Why is that? Because plants not only provide food for us to eat, they also release **oxygen** into the air for us to breathe. What's super interesting about this is that God created humans to breathe out carbon dioxide and breathe in oxygen. The stomata take in carbon dioxide and release oxygen. Oxygen is vital for human survival. Without the leaves that surround us on Earth, there would be less oxygen for you and me to breathe. In fact, about 20 to 30 percent of the world's oxygen comes from the tropical rainforests on Earth. The other 70 to 80 percent of our oxygen comes from plants in the ocean called phytoplankton. We should be super thankful for those green oxygen-supplying machines!

Here's another amazing thing about leaves. They actually clean the air! When I pulled the leaves off the ferns in my mother's house, I was not only removing the fern's food source; I was also removing our household air purifiers! The more plants you have in your house, the cleaner the air in your home. Plants clean the air by taking in carbon dioxide (that stuff we breathe out) and replacing it with oxygen. They also absorb harmful chemicals that are dangerous for us to inhale. Because plants absorb those chemicals, you and I do not breathe them. Instead, we breathe the wonderful oxygen the plants make for us. God is so creative and good to make plants that bless humans in such amazing ways!

Since you know that leaves produce oxygen, which plant do you think would produce the most oxygen—a plant with large leaves or one with small leaves? You're right! The bigger the leaves on the plant, the more oxygen it will produce.

MAKING FOOD

Making food is a big job for a plant. It takes a lot of complicated steps. Do you remember what kind of food plants like to eat? It starts with an "s" and is one of your favorite food ingredients, too. Plants and children (and adults) love sugar. Without sugar, plants would die. Where do you think the sugar is made? Take a guess! Sugar is made right inside the leaf! First, the leaf takes water from the roots. It then combines the water with light from the sun and carbon dioxide from the air to make sugar for the plant and oxygen for all the rest of God's creation. The plant sends the sugar down from the leaves, and the sugar runs throughout the entire plant to feed each and every part.

Have you ever broken the stem of a plant? Did you find gooey liquid dripping out? That liquid is called **sap**. It's a mixture of water, sugar, and other chemicals that travel throughout the plant. The sappy plant food gives the plant energy to make more leaves, to grow taller and stronger, and to make flowers. You remember how important flowers are to a plant, don't you?

Although this sap is yummy for the plant, it's usually not something we would enjoy eating. Sometimes it's even poisonous for us. However, some plants, like the maple tree, produce sap that's safe to eat.

Do you enjoy the sweet flavor of maple syrup on your pancakes? If you do, thank God for creating maple trees! Maple syrup comes from the sap of the maple tree. However, *real* maple syrup

is more expensive than the brand-name pancake syrups that line the grocery store shelves. That's because harvesting genuine pure maple syrup is not easy. Most brand-name syrups are just sugar water with fake maple flavor. Acquiring real maple syrup takes time. Let's find out how it works.

In the early spring, maple tree farmers poke holes into maple trees and place faucets in the holes. They then place buckets beneath the faucets. As the sap runs through the trees, some of it flows out of the faucets and into the buckets. The farmers then boil the sap and bottle it to send across the nation for people to buy and pour on their pancakes. Genuine maple syrup is very special because all of this must happen quickly. The sap is only sweet and tasty for about two to eight weeks. As a result, all of the sap used to make real maple syrup is collected in just a few weeks out of each year.

Aren't you glad God created the sap in maple trees to make your pancakes more tasty?

PHOTOSYNTHESIS

So how do the leaves make this sappy plant food? The process is called **photosynthesis** (foh' toh sin' thuh sis), and it's truly amazing. Each leaf is like a little sugar-making factory. When its stomata open, it starts taking in carbon dioxide. At the same time, water travels up from the roots of the plant to the leaves, and the leaf actually takes the carbon dioxide and combines it with the water. Sugar and oxygen are produced. The leaf uses some of the sugar for food then sends the rest of it down to the other parts of the plant, releasing the oxygen into the air for us to breathe.

SUNSHINE ENERGY

It turns out that the leaf can't actually do this job on its own. To combine the carbon dioxide and water, the leaf needs energy. Guess where it gets this energy? From light! That's the photo part of photosynthesis. You see, *photo* means light, and *synthesis* means to put together. So photosynthesis means putting together with light. That's exactly what leaves do. With the help of light, they put together carbon dioxide and water to make sugar and oxygen.

Now do you see why a leaf's stomata close at night? There's no light for photosynthesis, so the leaves can no longer make sugar and oxygen. If they can't make sugar and oxygen, they don't need carbon dioxide, so the stomata close, allowing the leaves to rest. This is good because the leaves need their rest! Seven days a week, they start working as soon as the sun rises and work nonstop, until the sun goes down. All of that time, the leaves are making sugar for the plant and oxygen for the rest of God's creation!

Without light, plants could not perform photosynthesis.

What have you learned so far about photosynthesis?

ACTIVITY 6.1
BURN A CANDLE IN A JAR

Fire uses oxygen as it burns and must have oxygen to keep burning. Let's do an experiment to see if a plant can give the fire enough oxygen to keep it burning.

You will need:
- 2 identical candles
- Matches or lighter
- Adult
- 2 large jars
- Small plant

You will do:
1. Go outside to a sunny spot.
2. Have an adult light both candles and place them a foot apart. Put the plant near one of the candles. Be sure it is not near enough to burn the plant.
3. You will need an adult to help you place the jars over the candles at the same time. Make sure the plant is inside one of the jars with the candle, as shown in the picture above.
4. When the oxygen inside each jar runs out, the candles will go out. Watch as the candles use the oxygen in the jar to burn.

Did one candle go out before the other? Why do you think that happened? Base your answer on what you learned about a plant's ability to produce oxygen. Why do you think I suggested that you do this experiment outside in a sunny spot? What does a leaf need to perform photosynthesis? The sun!

COLOR-FILL

Have you ever noticed that flowers come in many different shades and colors, but their leaves are mostly green? Why is this? The answer is they *need* to be green to do their job. Remember, leaves need light to make food for the plant. How do they use this light? Interestingly, leaves are filled with a special substance called **chlorophyll** (klor' uh fill). Chlorophyll absorbs the light the leaves need to do their job. Chlorophyll takes the light's energy and gives it to the leaf in just the right way so that the leaf can use it. And guess what? The chlorophyll is what makes the leaves green! We can remember this because *chlor* sounds sort of like color, and *phyll* sounds like fill. So chlorophyll fills the leaves with green color. Since chlorophyll is necessary for a leaf to use light for photosynthesis, you know that if a leaf is green, it can perform photosynthesis. If a leaf is not green (in the fall, for example), it cannot perform photosynthesis.

These fall leaves are unable to perform photosynthesis.

ACTIVITY 6.2

BLOCK THE SUN

If it's the time of year where the grass is still green in your area, you can see how a lack of sunlight will affect the color of the grass. Simply take a board and place it over a section of grass that's exposed to the sun. Keep checking under the board each day to see what happens to the color of the grass. Can you explain why this is happening? After a week, remove the board completely. How does the grass look now? Check the grass again in a few days to see the sunlight's effect.

You may have noticed from this activity that after covering your grass, it grew pale and maybe even lost all its color. Why? Because the board blocked the sunlight from the grass, and as a result, the grass could no longer perform photosynthesis. This means its chlorophyll was not being used, so the green color decayed away with the chlorophyll. What happened after you picked up the board? In just a few days, the grass regained its nice green color, didn't it? That's because picking the board up allowed sunlight to start hitting the grass again. The grass started making chlorophyll so that it could use the sunlight for photosynthesis.

Think about the seedling project from Lesson 2. You grew seedlings in a window, in a dark closet, and in a refrigerator. Do you remember that the seedling in the closet grew longer and taller than the other plants? This is because it was using all the stored food in the cotyledon to try to find a way out of the darkness and into the light. Did you notice it was paler in color than the plant next to the window? That's because it was not making much chlorophyll. Without light, it could not perform photosynthesis.

What does your family do with the food that's left over from dinner? Do you store it in containers? When a plant makes sugar through photosynthesis, it uses some of the sugar right away and stores the rest for later. Where does it store the sugar? Different plants store their food differently. Often, a plant will store its extra food in lumps that are in its stems or roots. A carrot plant, for example, stores its extra food in a lump in its root. We call that lump the carrot, and that's what we eat. A potato plant actually stores its extra food in an underground stem we call a tuber. A potato, then, is just a special underground stem in which the plant's extra food has been stored. Do you see how God takes care of us and the other consumers in creation? He designed plants to make more food than they need so they can store excess food for consumers to eat. It's amazing how God created plants to do this! Write your discoveries in your journal.

The potato plant stores its food in this underground stem we enjoy eating, called a potato.

Explain what you have learned so far.

ACTIVITY 6.3
SPROUT POTATOES

Have you ever noticed that sometimes the potatoes in your kitchen have little nobs on them? We call these nobs eyes. Typically, a potato won't grow these unless it is very old. This is because the farmer sprays a chemical on the potato to keep it from budding. However, if you get organic potatoes, you can grow an entire potato plant from a single potato tuber in only a few weeks. That's right. Even though the potato is actually a storage facility for the extra sugar, it can produce a whole new plant. Let's try to grow one!

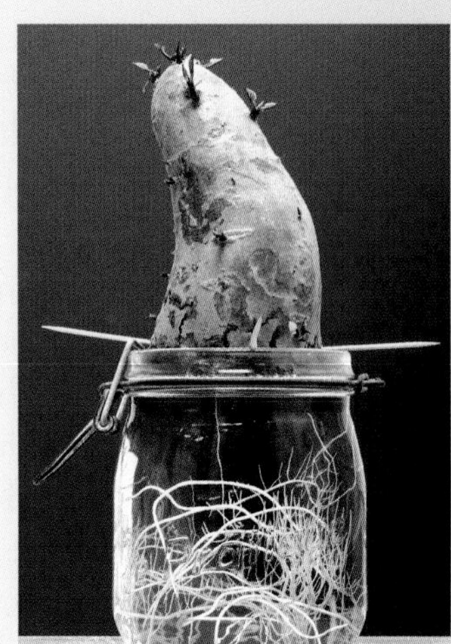

You will need:
- Large glass jar
- Organic potato (regular potatoes are sprayed with a substance to keep them from growing roots)
- 2 bamboo skewers
- Water

You will do:
1. Insert the bamboo skewers into the center of the potato.
2. Fill the jar with water.
3. Place the potato on the jar of water, balanced by the skewers, with half the potato extending below the water and half extending above the water.
4. Watch the potato each day and refill the jar with water as necessary. Keep in mind that certain varieties can take a few weeks to sprout.
5. Once your potato has roots and leaves, you can plant it outside to grow more potatoes. (Only plant it outside when there is no more danger of freezing weather).
6. Write the process of sprouting your potatoes and the results in your Botany Notebooking Journal.

TRANSPIRATION

Do you remember what the stomata do? They take in carbon dioxide from the air for photosynthesis. They also release oxygen from the leaves. But that's not all they do! Stomata have another important job. They release excess water from the plant that comes up from the roots. This is called **transpiration** (tran spuh ray' shun).

All day long, the plant's roots absorb water from the soil, sending the water through the plant's veins to the leaves. The leaves, in turn, use the water for photosynthesis. It turns out, however, that the roots send up more water than is needed for photosyn-

This South American rainforest is very humid because of the amount of transpiration taking place.

The large leaves in this rainforest take in and release large quantities of rainwater each day, contributing to the air's moisture.

thesis, so the leftover water is pushed out of the plant through the stomata. That's transpiration. As long as the plant is performing photosynthesis, it is also transpiring because it must get rid of the excess water.

Harlan One of the best places to study transpiration is in a rainforest. The rainforests are filled with many amazingly large plants that have gigantic leaves. The leaves grow large to better capture the very few rays of sunlight that peek through the dense trees of the forest. Because there's so much rain, the plants have many opportunities to transpire. They suck up the rainwater and release it through their leaves in large quantities each day. Because the leaves are so large, the air is filled with water, making it moist. When there's a lot of moisture in the air, it becomes very humid.

Dad The rainforest is a very humid place partly because of transpiration. There is so much water in the air that if a seed lands on a tree, it can begin growing right there on the tree because it has all the water it needs to germinate and keep growing! These types of plants are called *epiphytas*, from *epi* that means on top of and *phyta* that means plant—in other words "plants that live on top of other plants."

ACTIVITY 6.4

TEST TRANSPIRATION

Let's conduct an experiment to see transpiration occurring in a plant.

You will need:

- Plastic sandwich bag
- Clothespin
- Living plant that is not an evergreen

You will do:

1. Wrap the plastic bag around several of the plant's leaves (preferably broad, flat ones).
2. Seal the bag at the petiole (the little stalk that attaches the bottom of the leaf to the stem of the plant) with the clothespin.
3. Water the plant.
4. Observe the plastic bag several times a day for the next few days.

What do you think will happen? How long do you think it will take before you see results? Use a Scientific Speculation Sheet to record your hypothesis, experiment, and results.

FALLING LEAVES

What would happen to the plants in your yard if you pulled off all their leaves? Unless they had enough sugar stored inside them to survive long enough to make more leaves, they would probably die. As it turns out, in the winter, some plants shed their leaves and live off the sugar they stored up during the spring and summer.

So what makes a plant lose its leaves in the fall? Transpiration! You see, God designed some plants to lose their leaves as a way to protect the plant during the frozen, dry winter. Because rain is scarce and water is often frozen in winter, it's difficult for the roots to get enough water for the plant or tree. Do you remember what happens in transpiration? Water vapor escapes the plant through its leaves. What if the plant continued to lose water through the leaves all winter and couldn't replace it with water from the roots? The plant would die of thirst!

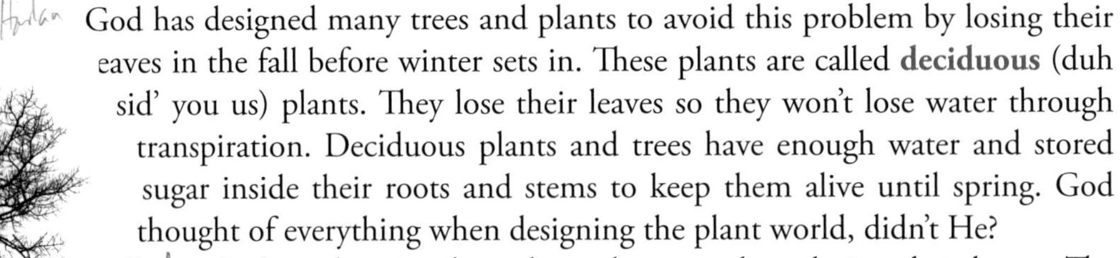

God has designed many trees and plants to avoid this problem by losing their eaves in the fall before winter sets in. These plants are called **deciduous** (duh sid' you us) plants. They lose their leaves so they won't lose water through transpiration. Deciduous plants and trees have enough water and stored sugar inside their roots and stems to keep them alive until spring. God thought of everything when designing the plant world, didn't He?

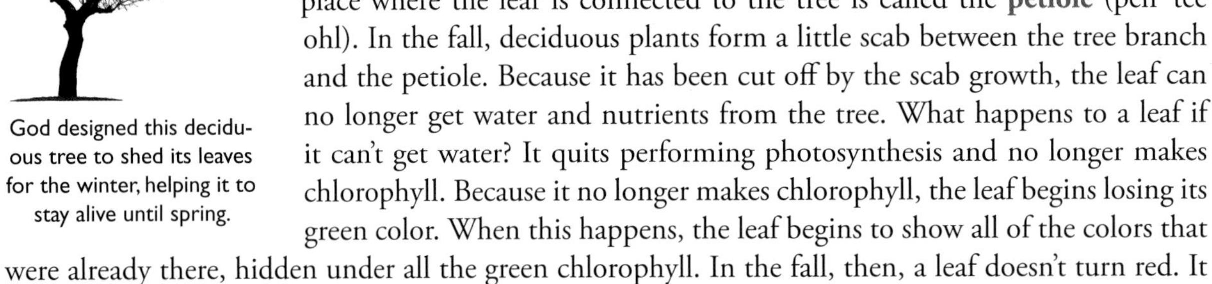

God designed this deciduous tree to shed its leaves for the winter, helping it to stay alive until spring.

Let's explore just how these plants go about losing their leaves. The place where the leaf is connected to the tree is called the **petiole** (peh' tee ohl). In the fall, deciduous plants form a little scab between the tree branch and the petiole. Because it has been cut off by the scab growth, the leaf can no longer get water and nutrients from the tree. What happens to a leaf if it can't get water? It quits performing photosynthesis and no longer makes chlorophyll. Because it no longer makes chlorophyll, the leaf begins losing its green color. When this happens, the leaf begins to show all of the colors that were already there, hidden under all the green chlorophyll. In the fall, then, a leaf doesn't turn red. It already was red! The green chlorophyll was so dark, however, that all spring and summer it covered up the other colors that were already there in the leaf. As the chlorophyll decays away and is not replaced by new chlorophyll, the beautiful colors that the chlorophyll was hiding show up. This time of year when the leaves begin turning their beautiful hues of red, yellow, and orange is one of the seasons people enjoy most. God is so good to create and share this beauty with us!

The green chlorophyll has faded away, revealing these leaves' beautiful colors underneath.

As the chlorophyll in the leaf decays away, the connection between the petiole and the tree gets weaker and weaker—so weak that when the wind blows, the leaf falls off the tree. This tells us that winter is on its way.

Describe transpiration in your own words. Explain why some trees lose their leaves in the fall and how the plant survives the winter without its leaves. Be sure to use the scientific name for these plants.

ACTIVITY 6.5
PRESERVE LEAF COLORS

If you're enjoying the fall season in your area, you can actually preserve the color of the leaves and make a wonderful decoration to hang on your walls.

You will need:
- Mod Podge®
- Card stock or blank canvas
- Fall leaves
- Paintbrush

You will do:
1. Go outside and collect the most beautiful fall leaves you can find.
2. Cover your paper or canvas with a layer of Mod Podge.
3. Arrange your leaves on the canvas or paper. With the paintbrush add a layer of Mod Podge on top of the leaves. You may want to repeat this process a few times until the leaves are completely flat on the page and covered well with the Mod Podge.
4. Once your creation is dry, you can frame it, add it to your nature journal, or place it in your Botany Notebooking Journal.

ANATOMY OF A LEAF

When we study **anatomy**, we're studying the individual parts that make up a larger body. For example, all the different parts of a leaf make up the leaf's anatomy. The drawing below shows you the basic anatomy of a simple leaf.

The very tip of the leaf is called the **apex** of the leaf. The petiole is the stalk that attaches the leaf to the plant, and the **midrib** is the main vein of the leaf. It's actually an extension of the petiole, and it runs most of the length of the leaf. If the leaf has branching veins, the other veins start at the midrib. The entire leaf above the petiole is the called the **lamina** (lam' ih nuh) or blade, and the edge that surrounds the leaf is called the **margin**.

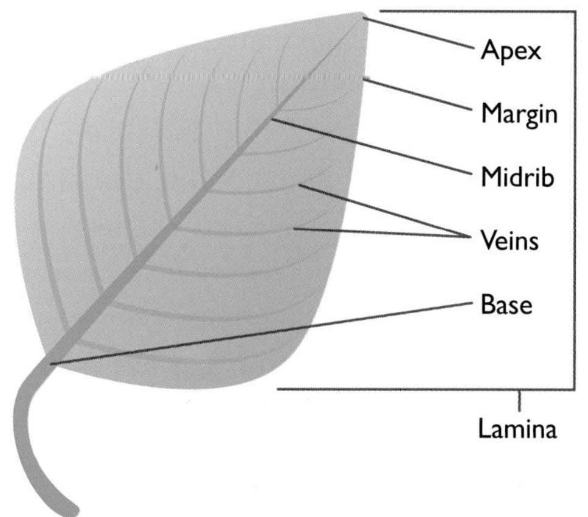

Now that you've learned the leaf's basic anatomy, find a leaf and see if you can name the parts without looking at the drawing. Why is it important to learn a leaf's anatomy? Because different plants have leaves with different types of margins, veins, and so on. If we can learn the different kinds of leaf parts in creation, we can identify plants based on their leaves.

SIMPLE LEAVES AND COMPOUND LEAVES

Leaves are often put into two main groups based on how they're attached to the plant. A **simple leaf** is one leaf attached to the plant's stem by a single petiole. A **compound leaf** has several leaflets attached to a single petiole. We call such leaves compound leaves because *compound* means more than one.

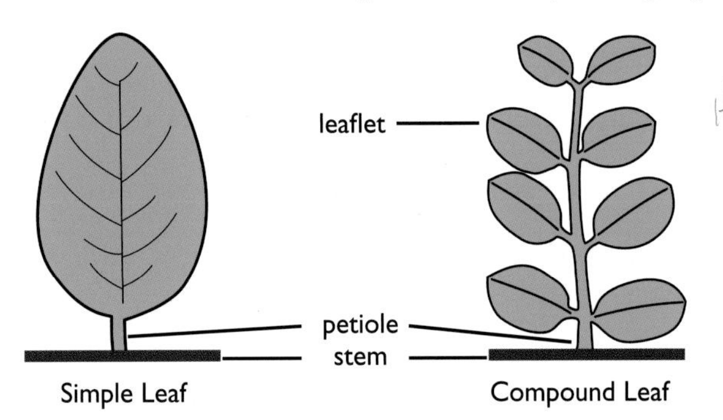

The drawing on the left illustrates the difference between simple and compound leaves. Although the compound leaf may look like a stem with several leaves on it, it's not. The stem of the plant is shown on the bottom, and a single petiole connects the entire compound leaf to the plant. The individual blades on a compound leaf are usually called **leaflets**.

It may take some time for you to get used to recognizing compound leaves. The key is to determine where the petiole attaches to the plant. If you can determine where the petiole is, you can then see if it leads to only one leaf (which means the leaf is a simple leaf) or many leaflets (which means the leaf is a compound leaf).

LEAF ARRANGEMENT

The arrangement of leaves on the stem often helps to identify a plant. There are three basic arrangements for leaves: **opposite**, **alternate**, and **whorled**. Leaves that are directly opposite one another going up the stem have opposite arrangement. Leaves that are not directly across from one another on the stem, but alternate as they travel up the stem, have alternate arrangement. Finally, leaves that are attached to the stem and arranged in a ring around the stem have whorled arrangement.

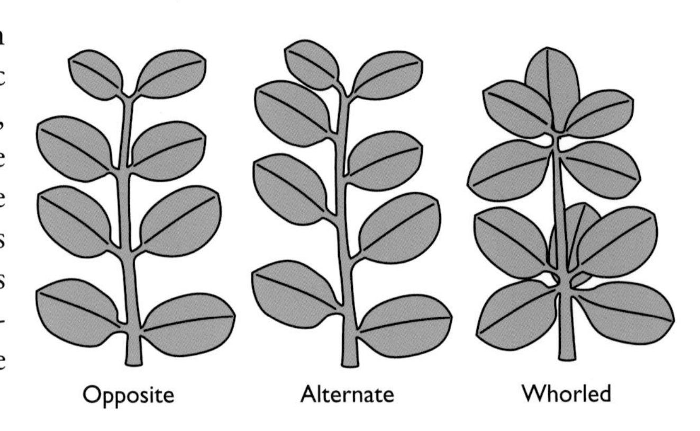

Parallel Pinnate Palmate

LEAF VENATION

As you've already learned, the pattern of veins on a leaf, which we call the leaf's **venation** (ven ay' shun), can tell you a lot about a plant. You learned in Lesson 2 that monocots produce leaves with veins that do not branch, while dicots produce leaves with veins that branch. When a leaf's veins run up and down the leaf without intersecting, we say it has **parallel venation.** And when a leaf's veins all branch out from a single vein in the middle (the midrib), we say the leaf has **pinnate venation.** Finally, the leaf has **palmate venation** when there are several large veins starting at the bottom (or base) of the leaf. These large veins have smaller veins branching off of them.

Now remember, monocots produce leaves whose veins don't branch. This means monocots have leaves with parallel venation. The leaves of dicots, however, have veins that branch. As a result, dicot leaves can have either pinnate or palmate venation.

LEAF SHAPES

As you already know, God created His world with a lot of variety. You've seen that plants produce many different kinds of flowers, fruits, and seeds. It's not surprising, then, that they also produce a wide variety of leaves. In this section, I want to teach you the different kinds of leaf shapes that we see in creation. This is not a complete list, but it does cover most of the leaf shapes you'll find in nature.

Before we explore these shapes, it's important for you to understand that you'll use this information for reference. In other words, when you need to determine the shape of a leaf, you can come back to this part of the book and look it up. For the leaf classification activity near the end of this lesson, you'll want to come back to these pages to compare the leaves you find with the drawings that are here. This will help you identify the shapes of the leaves you've found.

Many leaves in creation are a lot longer than they are wide. Sometimes the leaves are **linear**, which means they're about the same width from the top of the leaf to the bottom. If the leaf is wider at the bottom (near the petiole) and tapers toward the top (the apex), we say it has a **lanceolate** (lan' see uh late) shape. The upside-down version of the lanceolate shape is the **oblanceolate** (ob' lan' see uh late) shape. In this shape, the leaf is broader at the apex and tapers down to the petiole. If the leaf tapers at both the petiole and the apex, but is still longer than it is wide, it has an **elliptical** (ee lip' tik uhl) shape. If, on the other hand, there is no tapering on either end, and the leaf is about twice as long as it is wide, we say the leaf has an **oblong** shape.

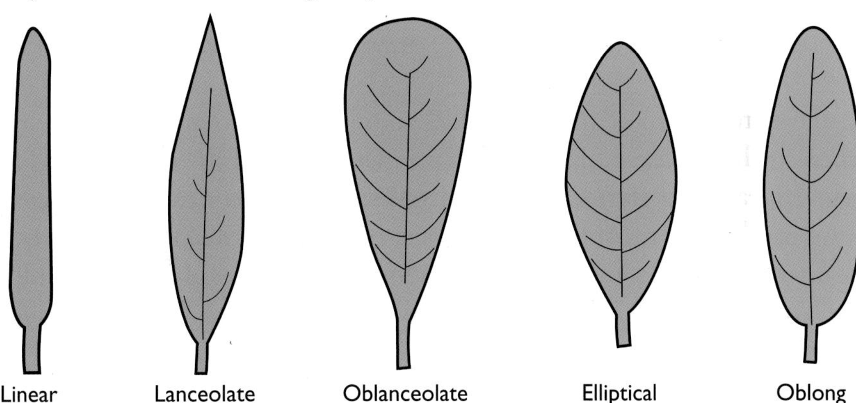

Although many leaves are a lot longer than they are wide, there are some that are more egg-shaped. In other words, they are only a bit longer than they are wide. If a leaf is egg shaped, we say it has an **oval** shape. If it's shaped a little like an egg but tapers toward the apex, its shape is called **ovate**. Now don't get confused between ovate and lanceolate. Both are wide at the bottom and taper toward the top, but a lanceolate leaf is a *lot* longer than it is wide, while an ovate leaf is only a *little* longer than it is wide. The upside-down version of the ovate shape is the **obovate** (ah boh' vayt) shape, which is egg-shaped but tapers toward the petiole. Finally, if the leaf is somewhat egg-shaped but the taper toward the petiole is very long, the leaf shape is **spatulate** (spat' you late), which means spoon-shaped.

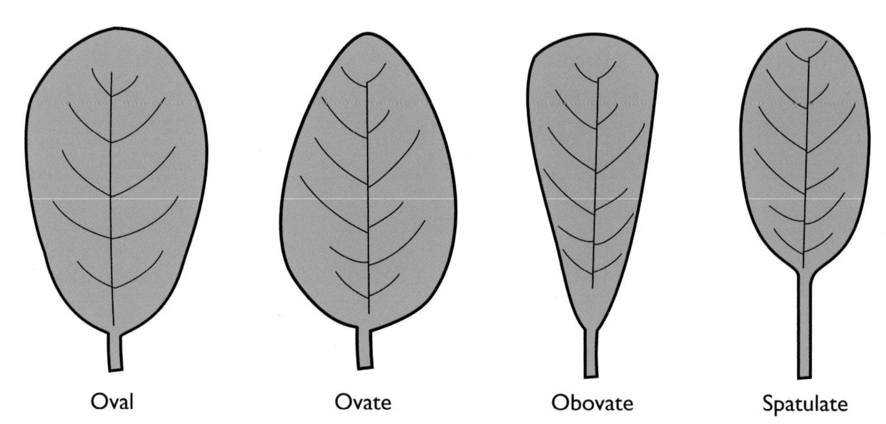

Some leaves have shapes that are very familiar to us. They often look like triangles, circles, or hearts. A leaf that is triangular in shape is called a **deltoid** (del' toyd) leaf. If it looks like an upside-down heart, it's a **cordate** (kor' dayt) leaf. If it looks a bit like an upside-down heart but is much wider and more circular, it is a **reniform** (which means kidney-shaped) leaf. If it's shaped more like a wedge, it's a **cuneate** (kyou' nee ate) leaf. If the leaf is nearly circular, we call its shape **orbicular** (or bik' you lur), which means circular.

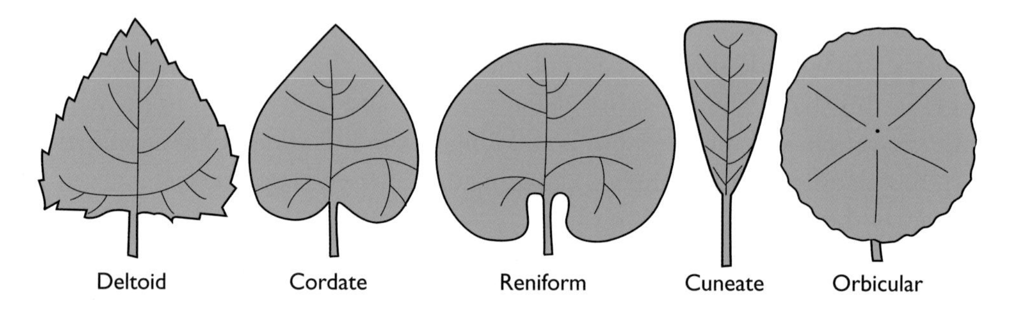

Finally, there are some leaf shapes that are more irregular than the ones I've discussed so far. For example, there are **lobed** leaves that have deep indentations. If the indentations are very deep and tend to be sharp, the leaf has a **cleft** shape. The needles on pine trees are actually leaves, and although they are linear, we usually give them their own shape, calling them **needle-like** leaves. Some trees have leaves that look a bit like needles, but they're not as long as needles and look almost like thin triangles. We say these are **awl-like** leaves. Finally, some plants have leaves that look almost like the scales on a fish. Not surprisingly, they're called **scale-like** leaves.

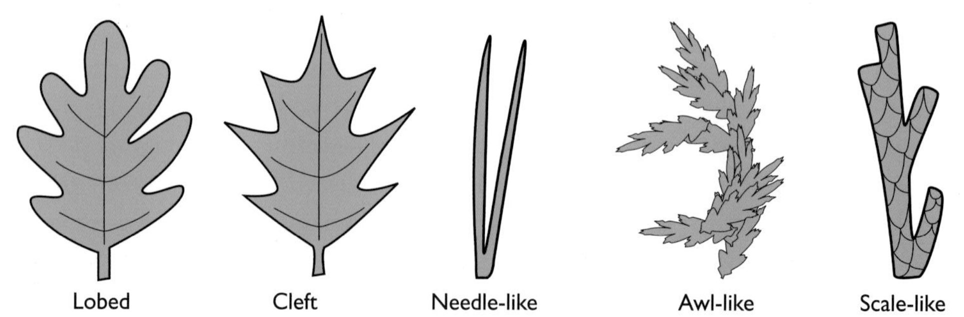

Now please remember these are not the only leaf shapes in creation. However, they do represent most of what you see when you're looking at leaves.

LEAF MARGINS

Do you remember what the margin of a leaf is? It's the outer edge of the leaf, and there are many different types of margins a leaf can have. If the outer edge of a leaf is smooth with no indentations or teeth, it has an **entire** margin. If, on the other hand, the leaf has tiny, sharp teeth along its outer edge, it has a **serrate** (seh' rate) margin. With serrate margins, the teeth usually point upward toward the apex of the leaf. If a leaf's outer edge has more pronounced teeth that also point outward rather than just toward the apex, it has a **dentate** margin. If the teeth are rounded rather than pointed, the margin is called **crenate** (kree' nate). Finally, if the leaf's edge does not have teeth but tends to be wavy, we say it has an **undulate** (un' joo late) margin.

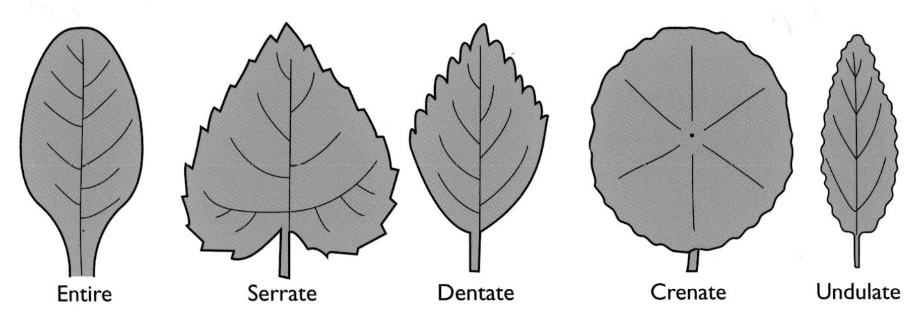

Now the real trick to identifying leaves is to try to determine the venation, shape, and margin all at the same time. You'll have plenty of opportunities to try this yourself when you do the second notebooking activity described in the next section.

You have learned so much about plants this year. You are truly becoming an expert! Before you begin the activities below, see how much you can remember from this lesson.

WHAT DO YOU REMEMBER?

Why are the leaves of a plant so important? Can you explain to someone what the stomata do for a plant? Explain what would happen if a plant lost all its leaves at once. What does a plant take in from the air and what does it put back into the air? Explain photosynthesis in your own words. What are the four things a plant needs to make food? What kind of food does the plant make? What happens when one ingredient is removed? What makes leaves green?

ACTIVITY 6.6

ILLUSTRATE THE ANATOMY OF A LEAF

Draw a leaf for your Botany Notebooking Journal labeling all of the parts of the anatomy. You can use the drawing on page 103 as a guide. Also, make a drawing that illustrates the difference between a simple leaf and a compound leaf.

ACTIVITY 6.7
MAKE A LEAF STORYBOOK

I want you to make a storybook about the life of a leaf. You can write the story yourself or dictate it to your parent or teacher to write down for you. Give your leaf a personality and a name. Start your story with the leaf beginning its life in the spring, going through the warm summer, and eventually falling off the tree to join his friends on the ground. Make sure you include the fact that the leaf works very hard to make food for the plant. Mention how its tiny mouths take in carbon dioxide and let out oxygen. Does the leaf enjoy this hard work, or does he resent it? Does the leaf like making oxygen for animals and people? When the leaf falls off the tree, does he enjoy being raked into a pile to be jumped on? Draw pictures for each phase of your leaf's life.

ACTIVITY 6.8
COLLECT LEAVES

Now it's time to add some spice to your Botany Notebooking Journal. You are going to collect leaves then classify each leaf according to its arrangement on the plant, its venation, its shape, and its margin. If you're able, go outside and pick one of each kind of leaf that you can find. Remember, it's okay to collect leaves from plants, but be careful to take only what you need. If you cannot find many leaves outside, go to a plant nursery and ask if you may remove one leaf from each plant.

After you've collected the leaves, look back through the information presented in the past few sections to see if you can classify each leaf by arrangement, venation, shape, and margin. Use a field guide to identify the plants from which your leaves came.

You can preserve your leaves by covering them with Mod Podge and hanging them to dry. After you've done that, tape or glue your leaves into your Notebooking Journal. Record next to each leaf what you know about it, such as its arrangement, venation, shape, and margin.

LESSON 7

ROOTS

digging deeper

In Matthew 13, Jesus tells us a story: "*A farmer went out to sow his seed. As he was scattering the seed, some fell along the path, and the birds came and ate it up. Some fell on rocky places, where it did not have much soil. It sprang up quickly, because the soil was shallow. But when the sun came up, the plants were scorched, and they withered because they had no root. Other seed fell among thorns, which grew up and choked the plants. Still other seed fell on good soil, where it produced a crop—a hundred, sixty or thirty times what was sown.*"
Matthew 13:3–8 (NIV)

Later, Jesus explains this verse to his disciples: "*When anyone hears the message about the kingdom and does not understand it, the evil one comes and snatches away what was sown in their heart. This is the seed sown along the path. The seed falling on rocky ground refers to someone who hears the word and at once receives it with joy. But since they have no root, they last only a short time. When trouble or persecution comes because of the word, they quickly fall away. The seed falling among the thorns refers to someone who hears the word, but the worries of this life and the deceitfulness of wealth choke the word, making it unfruitful. But the seed falling on good soil refers to someone who hears the word and understands it. This is the one who produces a crop, yielding a hundred, sixty or thirty times what was sown.*"
Matthew 13:19–23 (NIV)

The seed Jesus is referring to is the word of God, and the ground the seed falls on is our hearts. As children of God, let's make sure our hearts are the good soil that nourishes the word of God.

GOOD SOIL

So far, we've been focusing our attention on the seed and sugar factories you find above the ground in plants. Now let's dig below the surface to uncover the little water pipes beneath, the **roots**. A plant's root system is really a network of tubes, usually growing underground. The roots have two main jobs: to absorb nutrients and water from the soil and to hold the plant in place like an anchor.

How long do you think you could survive without water? Scientists say that, on average, a human can last only three or four days without water. A plant is much like people in that way. God designed plants to depend on water for survival. As you know, they can't go to the sink and get a drink of water like you and I can. Plants depend on rain and underground water for survival. That's because the main way a plant gets water is through its roots.

From the roots, the water goes up into the rest of the plant to be used in photosynthesis. The roots absorb other things besides water. For example, they absorb valuable nutrients from the soil.

But the plant doesn't use these nutrients for food. Remember, plants make their own food. However, the plant does use the nutrients in other ways to keep itself healthy. You can think of these nutrients as vitamins for the plant. Just as vitamins help keep you healthy, the nutrients that plants absorb from the soil help keep them healthy. These nutrients are so important to a healthy plant that a plant cannot produce healthy fruit without them.

This plant's roots will soak up the rain that seeps into the ground

NUTRITIOUS SOIL

Do you remember we talked about the fact that a plant can absorb all the nutrients in a plot of ground, leaving no nutrients for new plants? This can be a problem for farmers. If a farmer's soil is depleted of all its nutrients from the previous plants, what should the farmer do? In many cases, he adds nutrients back into the soil using manmade chemicals called fertilizer. Sometimes the farmers let the land rest for a year so it can replenish its nutrients. How can land replenish its supply of nutrients? Let's explore that for a moment.

As you know, air provides a few of the nutrients plants need, such as carbon dioxide. Three other very important nutrients plants must have are chemical elements called nitrogen, phosphorus, and potassium.

A plant cannot perform photosynthesis without the right amount of nitrogen. If your plants aren't very green, they might not have enough nitrogen. There's a great deal of nitrogen in Earth's atmosphere, but plants don't breathe it. They need a form of nitrogen that's fixed in the ground. Here's one way this happens: When plants and animals die, living organisms consume them. While they are consuming them they produce fixed nitrogen that plants can then take into their roots and use for photosynthesis. So you see, dead plants and dead animals are essential for replenishing nutrients into the soil.

Farmers provide nutrients to their crops by adding chemical fertilizers.

Phosphorus provides plants the energy they need to grow. If your plant isn't growing, it may need more phosphorus. Phosphorus isn't found in the air either, but it can be found in many rocks. When rocks erode (or break down), they add phosphorus to the ground. But that takes a long time. Interestingly, phosphorus is found in animals' teeth and bones. It's also found in plants because they must have it to grow. So decaying plants and animals (which have phosphorus in them) add phosphorus to the ground when they die. Manure, which is the droppings of plant-eating animals such as cattle and horses, contains phosphorous as well.

Potassium is the other super important element plants need for photosynthesis. It can be found in both rocks and living things. As with nitrogen and phosphorus, decaying plants and animals are important to replenish potassium in the soil.

Plants also need sulfur, magnesium, and calcium, but in smaller amounts. In fact, there are at least 18 known nutrients a plant needs to grow into a healthy, vibrant plant.

This farmer is spreading phosphorous rich manure to help his crops grow.

You'll learn in the gardening lesson how to ensure your plants have nutrient-rich soil without adding fertilizer. You'll learn to grow plants organically, meaning naturally, without a need for manmade chemicals. You'll do this by adding compost to the soil. Compost is a type of soil that's made up of decayed plant and animal matter. You can buy compost at the store, and there are many different kinds: cow manure, chicken manure, mushroom compost, earthworm castings, and the like. However, you can make your own compost by saving kitchen scraps, grass, dead leaves, and other plant matter from outside. Simply throw these items into a container in your yard and watch it turn into soil. I don't recommend adding dead animal parts to your compost pile as it would stink terribly and draw unwanted animals to your yard. It usually takes about two years for organic matter to completely decay into the compost soil you can use for your plants. However, you can make a small amount of compost in a short amount of time (only a month or two). Let's try it!

ACTIVITY 7.1

MAKE A QUICK COMPOST

Like I mentioned before, it usually takes a year or two for food and plant scraps to decompose and turn into compost soil. However, you can speed the process by chopping, turning, and mixing the compost regularly. In fact, you can make a quick compost in less than two months if you are super diligent. Would you like to try? Here's what you should do.

You will need:

- Black bucket with lid
- Adult
- Hammer and nail (or drill)
- Compostable material (see list that follows)

You will do:

1. Have an adult drill or hammer (using the nail) at least six holes in the bucket. This allows air to flow into the bucket and also microscopic decomposers to enter the compost.
2. Begin filling the bucket with kitchen and garden scraps (see list). For a quicker compost, cut them into small chunks and pieces before tossing them into the bucket. Put the lid on the bucket and place it in a sunny location to increase the temperature in the bucket (higher temperatures will increase decomposition). Add water if the compost mixture is too dry.
3. Once a day, shake and roll the bucket to mix up the compost. Every other day, use a shovel to chop the contents of the compost into smaller and smaller pieces to speed decomposition. Continue to chop, shake, and mix up the compost each day. Within a month or two, you will have rich compost dirt to use in your garden.
4. Be aware that within a week or so, the compost will begin to smell quite strongly. As it decomposes, the smell will become increasingly powerful. This is good! The smell means the compost is turning into nutritious soil!

THINGS TO ADD TO A COMPOST	THINGS NOT TO ADD TO A COMPOST
Fruits and vegetables	Fruit pits (they will grow into a new plant before they decompose)
Leaves	Meat (fish, chicken, beef, or bones)
Coffee grounds and filters	Bread
Tea leaves and bags	Oil
Flowers	Milk products
Hair	Cheese
Dirt	Leftovers
Newspaper	Animal waste or kitty litter
Eggshells	Styrofoam
Plant trimmings	Glass
Grass clippings	Diseased plants
Paper bags	Sawdust
	Walnuts
	Diapers
	Metal
	Plastic products

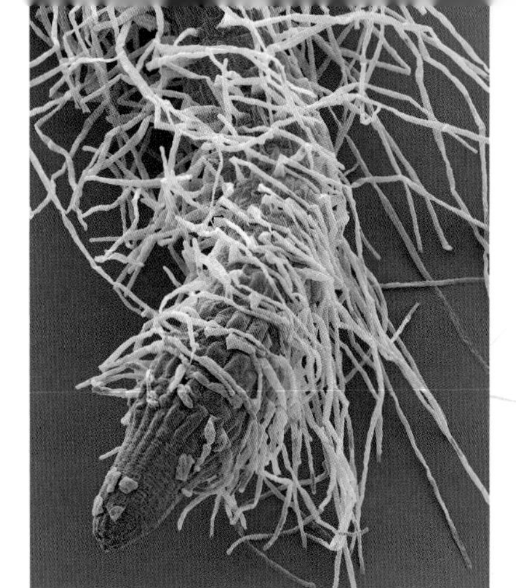

Root hair of a poppy seedling, viewed under a microscope.

ROOT HAIRS

Do you remember growing seeds in a bag? Did you notice that the surface of the roots was covered with tiny hairs that look a lot like fur? They appear soft and downy and seem a bit unimportant, yet they're probably the most important part of a root! You see, these hairs are called **root hairs**, and they actually do most of the work of absorbing water and nutrients for the plant!

Each little root hair lives for about six weeks, working mightily throughout that time to serve the rest of the plant. Root hairs lead a very short but noble life! After six weeks, new root hairs develop to take the place of the old ones that have withered and died. The new root hairs take over the job of drinking in water and nutrients for the plant. Without root hairs, a plant is likely to die.

This plant's roots have been preserved so it can be successfully transplanted to another location.

When people transplant (move to another location) a plant in their yard, it often dies because too many of the root hairs were destroyed when the plant was moved. Most people don't realize the most important parts of the roots are the root hairs. If enough root hairs are preserved when the plant is moved, the plant may be able to survive transplantation. Sometimes, however, the plant will look like it's dying before it can revive itself. This is often called **transplant shock**. If you're careful to dig a deep, wide circle around the plant, preserving the very tips of the roots and many root hairs, you can move a plant with little or no transplant shock.

ROOT GROWTH

Roots grow throughout the life of a plant. They grow longer from the tip, adding cells to the end of each root. Remember, cells are the basic building blocks of life. You might think that a root grows by adding cells to the base of the root, but that's not right. Roots grow longer by adding cells to their tips, and they grow fatter by adding cells around their tubelike bodies.

To better understand how roots grow, think about building a tower with Lego®. How would you do that? You'd probably begin building the base of the tower and then add Lego blocks to the top of the tower, making it taller. I bet you'd also add Lego around the sides, making the tower wider. This is how roots grow. Interestingly, you and I don't grow this way. Our bodies add cells in order to make them grow, but the cells are added everywhere, in a very complex way.

At the tip of each root, there's a little group of tough cells called the **root cap**. The root cap is the strongest part of the root tip, and its job is to push its way through the dirt to search for moisture and nutrients.

Just like when you add Lego to the top to make your tower taller, roots add cells to their tips to grow longer.

The root tip is so strong that as it grows it can push its way through cement and the concrete foundations upon which homes and buildings are laid. Tree roots can lift up whole houses over time! Have you ever walked along a sidewalk in an older neighborhood? Often, the sidewalks are full of cracks and are quite uneven. This is usually because of roots growing underneath the sidewalk, shoving the cement out of the way. When you see an uneven sidewalk, it's probably the fault of a root somewhere down below.

This cracked sidewalk is the result of the tree's root growth.

Look at the picture on the right. It's a picture of a root tip, but the picture was taken through a microscope. Remember, microscopes allow us to see things that are too small to see with our eyes alone. The squares and rectangles with little dots inside them are the cells of the root. Do you see the tiny root hairs in the picture and the area in which cells are being added to make the root grow longer? Notice how the root cap looks different from the rest of the root tip. That's because it's the strongest part of the root tip.

Roots can grow extremely long. You might be digging in your yard one day and run into roots from a nearby tree. The tree might not be as near as you think, however. You see, roots can grow a long, long way in search of water and nutrients. A wild fig tree in South Africa, for example, grew roots that were more than 393 feet long! That's a lot of root growth!

This plant's roots have stretched out to find water and nutrients hiding behind the wall.

Roots will grow on and on in their search for water and nutrients. If water and nutrients are present near the surface of the ground, the roots will not need to grow deep. Instead, they'll grow outward so they can cover a larger area. Roots are greedy little things, too. They'll try to take over as much ground as possible in hopes of soaking up all the water and nutrients they can. If a dry spell suddenly occurs, or if the soil near the surface of the ground runs out of nutrients, the roots will grow deep, searching for what they need.

PREVENTING EROSION

In addition to providing nutrients and water to the plant, the plant's root system acts like an anchor. It keeps the plant from washing away with the rain or blowing away with the wind. An added benefit is the roots keep the soil from washing away. Rain and wind can whisk soil away, changing the surrounding landscape. We call this **erosion** (ee roh' shun). Roots help keep erosion from

Roots like these hold soil in place, helping to prevent erosion.

happening by clinging to the soil, holding it in place. If you're worried about a hillside or other area eroding away, you can help prevent the erosion by planting seeds. As the seeds grow into plants, the roots that the plants grow will hold the soil in place. As you've just learned, roots have three important purposes in creation: they anchor the plant to the ground; they provide nutrients and water for the plant; and they help to prevent erosion. Three cheers for roots!

FLOATING ROOTS

Do you remember that I said roots are usually found underground? Well in some cases, that's not true. In the rainforest, a lot of water is present in the air. Do you remember why? In the previous lesson, you learned that the rainforest is full of plants with large leaves that are constantly performing transpiration. Because of the large amount of rain and the large amount of transpiration that occurs in a rainforest, the air is very humid. This means it contains a lot of water. Because of this, you'll often find rainforest plants growing right on trees, with their roots hanging down into the air or running into the moss that's also growing on the trees. They don't even need dirt; they have all they need floating around in the air or hidden away in the moss! If you ever visit a place where it rains a lot, such as Georgia or South Carolina, you'll find many places where ferns and other plants are growing on trees.

This rainforest orchid can grow on trees because the moist air provides enough water for its roots.

Believe it or not, some plants have roots that are both above and under the ground. I'm not talking about trees with roots that peek out above the soil from time to time. I'm talking about plants whose roots sometimes start growing underground and sometimes start growing above ground! Look at the banyan tree pictured to the left. It has a root system that exists underground but also has roots that start in its branches and grow down toward the ground. Theses roots not only absorb water and nutrients from the soil, they help support the long branches of the banyan tree. Because of this extra support, banyan tree branches can be very, very long. Consider, for example, this banyan tree in Lahaina, Maui. It was planted in 1873 by a man named William Owen Smith. Its branches have grown so long that this single tree covers a full square block in the city!

This famous banyan tree is a full square block because of its amazing root system.

GEOTROPISM

God's design for roots is truly amazing. Think for a moment about what happens when you plant a seed. Do you have to worry about planting the seed right side up? No, you don't. No matter how you lay the seed in the dirt, the roots know to grow down into the soil. How do they know which way to grow? God has given them a sense of which way is down. This special sense is called **geotropism** (jee' oh trohp' is uhm). *Geo* refers to the earth, and *trop* means turning, so geotropism tells us that roots are always turning toward earth.

What causes geotropism? Well plants have chemicals called **auxins** (awk' suns) that affect how they grow. Gravity causes different amounts of auxins at different places in the root. Because of this uneven distribution of auxins in the root, the root grows in the direction of gravity, which is toward the center of Earth. When plants are grown in the International Space Station's Botany Science Lab, the roots don't always grow downward. They sometimes grow in the same direction as the stem! Since things in the International Space Station don't experience gravity, we can conclude that when a plant doesn't have gravity to guide it, its roots no longer know which way to grow. Because of this, geotropism is often called **gravitropism** (grav' oh trohp' is uhm), which means turning in the direction of gravity.

ACTIVITY 7.2

DISCOVER GEOTROPISM

You will need:

- Glass jar
- 5–8 paper towels
- Water
- 5–10 bean seeds

You will do:

1. Wet a paper towel and stuff it into the jar.
2. Line the bean seeds on the inside wall of the jar so you can observe them as they sprout. Arrange them sideways, upside down, right side up, and in all different directions.
3. Wet another paper towel and stuff it into the jar, lining it with more bean seeds.
4. Continue to wet paper towels and line the jar with seeds until you have filled the jar to the top.
5. Put the jar in a sunny window.
6. Keep the paper towels moist every day.
7. Watch how the bean seeds grow.

Which way do the roots turn if they begin growing the wrong way? What happens to the stems? We'll learn about stems in the next lesson. In your Botany Notebooking Journal, draw what your beans look like inside the jar once they have grown roots and stems.

ROOT SYSTEMS

As you continue to watch the roots from your bean activity grow, let's learn more about the different kinds of roots. There are two main kinds of root systems: **taproot systems** and **fibrous root systems**. A taproot system is made up of one thick, main root growing down from the plant's stem. It has a number of smaller **secondary roots** branching off from the main root, but a taproot system is deeper than it is wide. Often we eat taproots, such as carrots and turnips. Do you remember monocots and dicots from Lesson 2? Well monocots typically do not have taproots. If a plant has a taproot, it's most likely a dicot.

Beets are examples of taproots we enjoy eating.

A fibrous root system is made up of a series of roots growing in many directions. There may be a few roots that are thicker than others, but there is not one main root. A fibrous root system is usually wider than it is deep. Monocots typically have fibrous root systems, but not all fibrous root systems are from monocots. Some dicots also have fibrous root systems. As a result, leaves and flowers are the most accurate indicator when determining whether or not a plant is a monocot. Do you remember the distinguishing features of monocot leaves and flowers?

Fibrous roots are a system of roots growing in all different directions.

GEOPHYTES

Do you realize that roots are not the only plant parts you can find underground? That's right. Onions, garlic, tulips, and many other plants grow from little storage tanks called **bulbs**. Many people think these bulbs are part of the root systems of the plants, but they're not. Roots grow out of the bottom of a bulb, but the bulb is actually a special storage house that's made up of special leaves and a short stem. A bulb operates kind of like a seed since you can pull it up and save it to plant later. Bulbs are called **geophytes** (jee' oh fites). *Geo* means earth, as you previously learned, and *phyt* means plant. I guess they're called earth plants because they're like an entire plant under the eearth. Bulbs are not the only geophytes. There are other underground storage systems in plants such as corms, tubers, and rhizomes (rye' zohms). Each can be kept in a cool place for years then planted in the ground to grow. This is because each of these geophytes has all the nutrients the plant needs to grow its roots and stems. That sure does remind us of seeds, doesn't it?

A bulb looks a bit like a taproot, but each layer of a bulb is actually a special kind of leaf. Think about an onion. If you peel back the layers of an onion, you're actually peeling back leaves.

Onions and garlic are examples of bulbs.

A bulb is a source of constant renewal for a plant. In the winter, the plant's above-ground portion dies, but the bulb continues to live below the surface. When spring comes, the bulb begins growing new stems and roots, and the plant grows back above the ground again.

Rhizomes look like roots, but they're actually underground stems that store excess food. Instead of growing down like roots, they grow horizontally. If rhizomes are underground and look like roots, how do we know they aren't roots? Remember, roots absorb water and nutrients for the rest of the plant. Rhizomes don't do this, so they aren't roots. Roots can grow from rhizomes, however, as can leaves, flowers, and stems.

Corms are sometimes thought to be bulbs, but they're not. Bulbs have layers and layers of leaves, whereas corms are solid. They don't have layers that you can peel back. Corms grow roots from the bottom. However, just like bulbs, they survive underground during winter and produce new plants in spring. A crocus, often the first flower of spring, produces a corm.

These crocus corms will survive underground in the winter and produce a flower above ground in the spring.

Tubers are underground stems that are swollen, forming a big lump. Tubers are packed full of starch, which is a form of sugar. Where does the plant get all this sugary starch? Well, the plant turns the extra sugar it makes into starch because it's easier to store that way. Each tuber also has buds that can grow roots and stems. The most common tuber has buds we sometimes call eyes. This tuber is eaten every day by people all around the world. Can you guess what it is? A potato! As you can see, many geophytes are quite yummy!

Potatoes are delicious geophytes enjoyed by people all over the world!

ROOTING

Most plants grow from seeds; some plants grow from geophytes. There's another way to grow a plant, however. It's called **rooting** a plant. To root a plant is to take a healthy stem or branch from the plant and put it in soil or water. Within a few weeks, something incredible happens. New roots begin to grow from the branch! With some plants, you can root a new plant using just a leaf from the original plant. Some plants that root easily are: willow trees, African violets, ivy, geraniums, begonias, coleus, and roses. If you have any of these plants in your yard or neighborhood, try to root them in a vase of water. When you root a plant in water, you need to remove the bottom leaves from the stem and submerge the stem in the water. After a little time, roots will emerge from the stem, usually from where you pulled the leaves off. Isn't that amazing? If you want to root a plant in soil rather than water, place a cut leaf, root, or stem in the soil and keep the soil moist. It can develop roots and grow into a whole new plant from that one cutting!

This cut of rosemary is rooting in water in order to grow roots and become a new rosemary plant.

Harleigh Interestingly, a plant grown by rooting is different from a plant grown from a seed. Plants grown by rooting are called **clones**. What's a clone? A clone is a plant that has the same DNA as the plant from which the stem or root came. Do you know what DNA is? DNA is something every living thing (including you) has! It contains all the information that's needed for the living thing to grow and survive. For example, your DNA contains all the information your body needs to eat, sleep, talk, or do anything else that you do. In other words, DNA is the blueprint your body uses to become you! Your DNA is a combination of your mother's DNA and your father's DNA. A seed's DNA is the same. Not so with a clone. A clone has the same exact DNA as the original. When a plant grows by rooting, then, its DNA is the same as the plant from which the root or stem was taken. It's a copy of the original plant. This is not the case with a plant that grows from a seed.

Hart Has anyone ever told you that you have your father's eyes or perhaps your mother's hair? What did that person mean? He meant that your eyes look a lot like your father's eyes, or your hair looks a lot like your mother's hair. Most likely, you have some things in common with each of your parents. However, you don't look *exactly* like either parent, do you? That's because, as I told you already, your DNA is a mixture of your mother's DNA and your father's DNA. This is what it's like for a plant that grows from a seed. The plant has a mixture of the DNA from the parent that produced the pollen and the DNA from the parent that received the pollen and made the seed. As a result, the new plant has things in common with both its parent plants but is not a copy of either one of them. A plant that grows from a rooting, however, is really a copy of the original plant because it has the same DNA as the original.

Let's take a minute to remember everything we learned in this lesson.

WHAT DO YOU REMEMBER?

What are the three main purposes for roots? Explain why root hairs are important. What is the root cap? Where do roots add to their length? What are the roots always looking for? What is geotropism? What is another name for geotropism? Tell me about rooting a plant.

ACTIVITY 7.3
ILLUSTRATE ROOTS

In your Botany Notebooking Journal, illustrate some of roots you learned about. Record all that you remember about roots.

ACTIVITY 7.4

CLASSIFY ROOTS

Do you have weeds in your yard? If you do, today will be a fun day for you because you are going on a root classification hunt! Begin by dressing in clothing that can get dirty then put on some gloves. Garden gloves are best, but any gloves that protect your hands will work. Go outside and begin pulling up weeds carefully, so as to preserve the roots. You may need to use a trowel or a metal spoon to dig around the roots so you can keep them intact, or whole, as you pull them up.

In your Botany Notebooking Journal, sketch the different weeds you find and their root systems. Note whether the root system is a taproot system or a fibrous root system. You might also want to do some research to find out the name of each kind of weed you find.

ACTIVITY 7.5

CLONE VEGETABLES THROUGH ROOTING

Let's grow some vegetables from vegetables! I've included two green vegetables below, but you can also learn how to root onions and garlic by going to the link provided in the course website information at the beginning of this book.

You will need:

- Romaine lettuce scrap (the bottom two inches of the base)
- Celery scrap (the bottom two inches of the base)
- Glasses or jars
- Water
- Planter big enough for two vegetables
- Garden soil

You will do:

1. After cutting the tops off, place the vegetable scraps in their own jars.
2. Fill the jars with a half inch or so of water to cover the bottoms of the scraps. Change the water every day.
3. Place the jars in sunlight for 5–7 days or until they begin to sprout.
4. Plant the vegetables in the soil in your planter.
5. Water the vegetables in the planter until they grow big enough to harvest and eat. Enjoy!

You can also clone other lettuce varieties, green onions, parsley, cilantro, or almost any other vegetable that has a base still attached from which the stems grow.

LESSON 8

STEMS

digging deeper

For just as rain and snow fall from heaven and do not return there without saturating the earth and making it germinate and sprout, and providing seed to sow and food to eat, so my word that comes from my mouth will not return to me empty, but it will accomplish what I please and will prosper in what I send it to do." Isaiah 55:10–11

In this amazing verse, we learn that the word of God is like the rain that falls to make seeds sprout and bear fruit. God's word is written in the Bible. As you learn and pray God's word, you will find that it has a great deal of power.

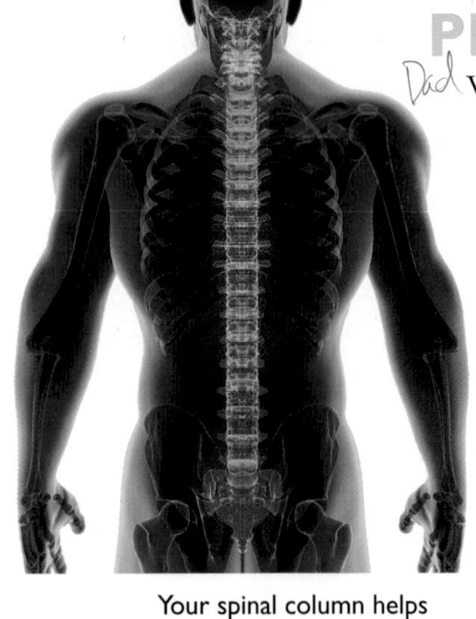

Your spinal column helps you stand tall.

PLANT STRUCTURE

What do you think gives you your shape? Could it be your skin? What about your muscles? These important parts of your body help, but there's another part that is largely responsible for your body's shape. Have another person stand up and bend over from the waist. Feel the ridge along his or her back. You have just felt that person's spinal column! The spinal column is the set of bones that helps us stand up straight and tall. Our bones are a major part of what gives us our form. What would happen to you if all your bones were removed? Why, you would fall into a heap upon the floor! Can you imagine that? How would you move about? Perhaps you would creep along the floor like a slug or a snail, struggling to move.

Believe it or not, plants also need to move. Not as much as you and I, but they do need to move. They also need structures that give them form. Most of a plant's movement and structure comes from its stems. Without stems, for example, plants would not be able to lift their lovely bodies above the ground. The stem is sort of like the plant's spinal column, giving it its form.

A plant's stem keeps the plant upright, just like your spinal column does for you!

There are many different kinds of stems, so there are many different plant shapes and forms. Can you describe a tree stem? Tree stems are hard and woody, aren't they? What about the stem of the foxglove plant pictured to the left. Is it hard and woody? No, it's not. It is soft and green. You might be surprised to learn that even grass has stems. Because people mow their lawns, we often don't see what full-grown grass looks like. However, if you look at the photograph on the on the next page, you'll see that the grass in the picture has a stem. Grass also has stems that grow sideways along the ground. These stems are called **runners**. Many plants produce runners. If you ever visit a farm that grows strawberries, look

A foxglove's stem is soft and green.

at the strawberry plants. They have stems that grow upward, and they also have runners that grow along the ground.

Harlan Some plants have woody stems, while others have hollow stems. Some stems are as short as a clover stalk, and some are as long as a vine. Although there is a lot of variety in plant stems, one thing is for sure: all vascular plants have stems of one kind or another. Do you remember what a vascular plant is? It's a plant with tubes that carry fluid throughout the plant. What exactly runs through these tubes? That's right! Water and nutrients! But there's something else that runs through the tubes. Do you remember what that is? It's the sugary food the leaves make for the plant. Water and nutrients run up from the roots to the leaves, and the sugary food comes down from the leaves to the rest of the plant.

Just like you have arteries and veins in your body that do different jobs, there are two different kinds of tubes inside vascular plants, and they do different jobs as well. The tubes that send water up from the roots to the rest of the plant are called **xylem** (zy' lum). And the tubes that send the sugary food down from the leaves to the rest of the plant are called **phloem** (floh' ehm). You can remember which is which because *phloem* sounds like flow, which is what happens to the sugary food when it flows down to be used by the rest of the plant. Inside every stem are bundles of these xylem and phloem. These bundles are called **vascular bundles**.

Can you explain in your own words the difference between xylem and phloem?

ACTIVITY 8.1

EXPLORE XYLEM

Let's do an experiment to see a plant's xylem in action!

You will need:

- Celery
- Cup
- Water
- Blue food coloring

You will do:

1. Fill the cup with water.
2. Add food coloring to the water, enough to make a deep blue color.
3. Put a stalk of celery in the cup of blue water.
4. In a few hours, take the celery out of the colored water.

Look at the bottom of the celery. Do you see the little dots? Those are the xylem that pull water up to the leaves at the top of the celery plant. Placing the celery in colored water helps you see how the xylem pull water up to the leaves.

WOODY AND HERBACEOUS STEMS

Biologists typically classify a plant as either **woody** or **herbaceous** (her bay' shus) based on its stems. If a plant has stems that are hard and woody, it is, not surprisingly, called a woody plant. If the plant's stems are green and easy to bend, it is called an herbaceous plant. It turns out that the stems of herbaceous plants and woody plants are quite different.

This orchid's herbaceous stem is green because it performs a lot of photosynthesis.

Do you remember what the plant's sugar-making process is called? It's called photosynthesis. What color tells you that a plant is making sugar through photosynthesis? Remember, leaves are green because they have chlorophyll, which is necessary for photosynthesis. Have you noticed that the stems of some plants are green? What do you think that means? It means that the stems of these plants perform photosynthesis. Green stems, then, not only support a plant and give the plant its form, they also help feed the plant.

The soft, green stems of herbaceous plants perform a lot of photosynthesis, while woody stems do not. Another important difference between these two types of stems is how they grow. In general, herbaceous stems do not grow thicker with time. They only grow longer by adding cells to the end of the stem. Woody stems, however, grow thicker and thicker as the plant gets older and older.

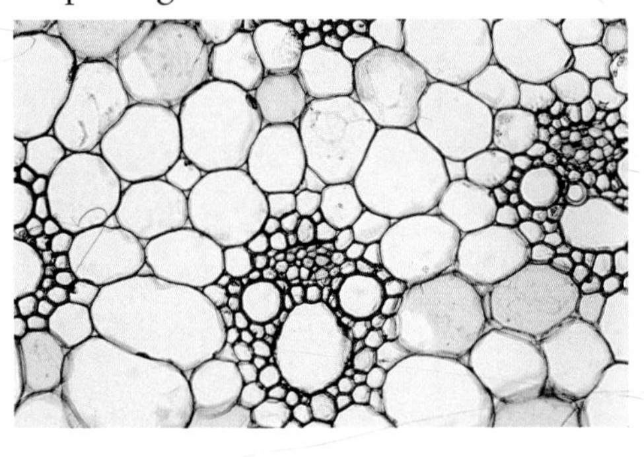

Woody stems grow thicker with age because they have something that herbaceous stems do not have. They have cells that make up a **vascular cambium** (cam' bee uhm). These cells are in a layer just inside the surface of the stem, as shown in the image on to the right. They form xylem on the inside of the layer and phloem on the outside. This thickens the stem. As time goes on, the thickness of the stem continues to increase as the vascular cambium makes more xylem and phloem. In other words, the cambium continues to add more and more wood to the stem, making it thicker and thicker.

There is something very interesting about the way woody stems grow in thickness. You see, the vascular cambium makes a lot of new wood during the spring and summer. As fall comes, it still makes new wood, but not nearly as much. In addition, the wood that it makes during fall looks different from the wood it makes in the spring and summer. When winter comes, the vascular cambium stops making new wood, waiting until spring to begin again. Once spring comes, it starts making lots of new wood, and that new wood looks different from the wood that was made during the fall.

This flower's woody stem looks very different from the orchid's herbaceous stem.

What does all this mean? Well, look at the picture of the tree trunk on the right. A tree trunk is the biggest stem on a tree. What do you notice about the trunk? There are a bunch of rings in the wood. The thick parts of the rings are made up of wood that was formed in

If we counted the rings in this tree's trunk, we could find out the age of the tree!

the spring and summer, and the edges of the rings are made up of wood that was formed in the fall. So each year, the stem forms a ring of wood you can actually see if you look at the inside of the stem. Each ring represents one year of growth!

Now think about what you just learned. Because of the way a woody stem thickens, it forms a new ring of wood each year. Because the inside of the ring is different in color than the outside of the ring, you can actually tell one ring from another. Suppose you see a tree that has been cut down. If you look at the inside of its trunk, you should be able to see rings, just as you saw in the picture on the previous page. Suppose you count those rings. What would you know? You would know how many years the tree had lived before it was cut down! Yes indeed, God has given us a way to tell how old a tree is! All we have to do is look at the inside of its trunk and count the rings. The next time you see a tree that has been cut down, try to count the rings on its trunk. That will tell you the age of the tree before it was cut!

These tree carvings will remain in the same spot on the trunk because the tree will grow taller from the top, not the bottom.

Now remember, the kind of growth I have been telling you about is how woody stems get thicker. Stems get longer in an entirely different way. Just like roots, stems grow longer by adding cells to their ends. The stem actually builds on top of itself in order to grow longer. Suppose you hammered a long nail into a tree two feet above the ground. In two hundred years, how high would that nail be? It would still be two feet above the ground. The nail might be deep inside the tree because the tree trunk got thicker every year, but it wouldn't move up as the tree grew because the tree does not grow from the bottom.

Suppose that many years ago, people carved their initials into trees. Those initials would be in the same spot many years later when they came back to the tree, even though the tree had grown taller. This is because the tree grows taller by adding cells to its top. Eventually the initials might be covered up by the new bark that forms on the outside of the tree, but they will not rise as the tree grows because as you've already learned a tree does not grow from the bottom.

ACTIVITY 8.2

DRAW WOODY AND HERBACEOUS STEMMED PLANTS

Go outside with your nature journal and look for the two types of stemmed plants you just learned about. Make a page with drawings of the woody-stemmed plants you find and a page with drawings of the herbaceous-stemmed plants. Be sure to use colored pencils to show the different colors of the stems. If you can identify the plant, write down what kind of plant it is.

Succulent plants have unusual and beautiful stems that look like leaves.

SUCCULENT PLANTS

Some of the strangest stems can be found in a group of plants call succulents. With succulents, part of the plant is very fleshy and thick. That's because God designed succulents to store a lot of the water they need to survive during long periods of drought. They are designed to have more areas for storage and less areas for transpiration. Succulents can be both dicotyledons and monocotyledons.

AUXINS

Do you remember what a seed needs to grow? It only needs water, air, and warmth to germinate. However, once it has used all the food within the cotyledon, the plant must produce food for itself. You already know it does this through photosynthesis. And one of the ingredients for photosynthesis to occur is light.

Light is so important to a plant's growth that God created its stems with special chemicals called auxins. In the previous lesson, I told you about auxins and how they cause geotropism in plants. Now I want to tell you a little more about auxins and how they also cause phototropism (foh' toh trohp' iz uhm) in plants. Do you remember what geotropism means? It means turning toward earth. You already know that photo refers to light, so what do you think phototropism means? Yes! It means turning toward the light.

Trees will lean as far as they need to find the light.

As it turns out, auxins make the stem of a plant bend and twist in order to reach toward the light. This is how God designed all plants to receive light as they grow. When something blocks light from a plant, auxins will cause the stem to bend, twist, and turn its way around the object to get to the sunshine. This is why we sometimes see trees with strangely curved and circled branches when we go hiking. Some have even bent into entire circles around other trees to reach the sunlight they need.

Scientists don't know all the details of how auxins cause phototropism, but here's how they think it works: It seems that auxins are destroyed by light. As a result, there are very few auxins in the places where light is hitting a plant's stem. However, in places where very little or no light is hitting the stem, there are a lot of auxins. Botanists know that auxins cause the cells of stems to stretch out. This causes the stem to stretch wherever its cells stretch. This means the parts of the stem that have little or no light hitting them will stretch, and the parts of the stem that have a lot of light hitting them will not stretch. This causes the stem to turn in the direction of the light. Later, we will do an experiment with bean plants to see auxins in action. In the meantime, try this activity to understand the process.

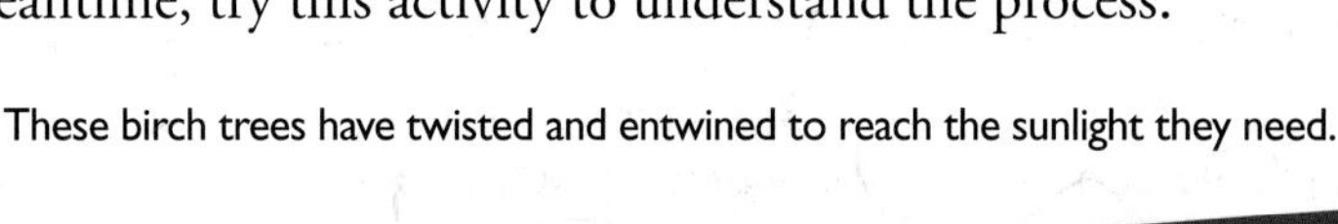

These birch trees have twisted and entwined to reach the sunlight they need.

ACTIVITY 8.3

IMITATE PHOTOTROPISM

Let's learn more about phototropism and the way auxins cause stems to bend so they can grow toward the light.

You will need:

- Clay or Play-Doh®

You will do:

1. Roll the clay into a fat "snake" and stand it on a table. This represents the stem of a plant.
2. Pretend that one side of the stem is getting plenty of light but the other side is in darkness.
3. On whichever side you chose as the dark side, pinch the clay with your thumb and forefinger and pull straight up on the clay. This simulates what auxins do. They cause the stem to stretch on the dark side of the stem.

What happened to the stem when you pulled it up? It bent toward the side that you assumed was getting plenty of light. By causing the side of a stem in darkness to stretch, auxins bend the stem toward the light.

EXAMPLES OF PHOTOTROPISM

Although all plants perform phototropism, there are some that really seem to love the light. As a result, they perform extreme phototropism. Consider sunflowers, for example. They are extremely phototropic and can turn toward the light rather quickly. They literally turn their heads all day long to face the sun. Every sunflower in a field will face the sun all day long because each flower twists its stems from the east to the west, following the sun as it travels across the sky. There is no mystery why they are called sunflowers, is there?

This phototropic sunflower twists its stem to face the light as the sun travels across the sky.

Daisies are another example of phototropism. The flower opens when the sun comes up and closes when the sun goes down. That is why it is called a "day's eye," or daisy.

In the summer, you may notice that plants grown in the shade are taller than plants grown in full sun. Once again, this is because of phototropism. Plants in the sun have plenty of light, so they tend to grow more outward than upward. Plants in the shade, however, are trying to find the light, so they tend to grow straight up. Remember the bean seeds experiment you did in Lesson 2? The seed that germinated in the dark grew taller than the seed near the window. Next time you are observing plants in nature, look to see if some plants are taller than other plants of the same kind. If so, are they shaded under a tree or bush, making their stems grow longer? Notice that the big and bushy plants are usually in full sun.

As the sun goes down, daisies fold in their petals to close for the night.

WHAT DO YOU REMEMBER?

What two parts of the plant are in the vascular bundle? Explain what xylem and phloem do for the plant. What is the difference between woody and herbaceous stems? What is phototropism? What are the special chemicals called that enable a plant to grow toward the light?

ACTIVITY 8.4

COLOR A FLOWER

Did you know you can color flowers? Let's try it!

You will need:

- Yellow, green, red, and blue food coloring
- 4 tall glasses
- 4 white roses or white carnations (If neither of these flowers are available at your local store, you can try to use another variety of white flower.)
- Water

You will do:

1. Fill each glass full of water.
2. Add a different color of food coloring to each glass.
3. Place a single rose or carnation in each glass. (Cut the stem before placing the flower in the water. The shorter the stem, the faster the results.)
4. Watch what happens.

Can you explain why this happened?

ACTIVITY 8.5

SEE AUXINS IN ACTION

You will need:

- 2 paper or Styrofoam® cups with a lid
- 4 bean seeds
- Mixture of vermiculite and compost for soil
- Sharpened pencil
- Black paint

You will do:

1. Paint or color one cup and the lid black.
2. Using a pencil, make a hole in each cup a half inch below the top of the cup.
3. Put three inches of soil in both cups.
4. Plant two bean seeds inside each cup.
5. Water both cups. (Be sure to make the soil wet, but not soaked. The seeds need some air.)
6. Put the black lid on the black cup. Do not cover the other cup.
7. Place the cups in your light hut or on a sunny windowsill. Water your plants regularly to keep them moist but not soaked. Try not to disturb the plants too much when you water them.

What do you think will happen? Why do you think this? How long do you think it will take for the results to be evident? Use a Scientific Speculation Sheet to record your hypothesis, what you did, and the results.

LESSON 9

GARDENING

digging deeper

Then God said, "I give you every seed-bearing plant on the face of the whole earth and every tree that has fruit with seed in it. They will be yours for food."
Genesis 1:29 (NIV)

From the very beginning of time, God desired that man should eat from his own garden. His original design was that we should actually live in a garden. Perhaps that's why gardening brings people such feelings of joy and accomplishment. God has many wonderful plans for your life, and one of them could very well be to make gardening a lifelong hobby!

YOUR EDIBLE GARDEN

Now that you've learned so much about plants, it's time to put that knowledge into practical use by growing your own food. Because you know what plants need to grow and be healthy, and what the soil needs to grow strong plants, you can become an expert gardener! In this lesson, you'll be creating your own comestible garden. **Comestible** is another word for edible. That's because the Latin word *comestus* means to eat. So everything you grow, you can eat!

Everyone's outdoor area is different. Some people have lots of land and sun, while some have mostly trees and shade. Others have only a small patio. Some people live in a warm climate with a long growing season, while others live in a cool climate with a short growing season. Because of this, I can only offer ideas for making your garden. You'll have to determine the best way to build and grow your garden based on your unique outdoor area.

My home is surrounded by trees. The only sunny area is on a small deck outside. Therefore, I created a mobile box in which to plant my garden. I'll show pictures of how I've done it, but remember, this is what worked best for me because I have no other way to provide the sunlight my plants need. Your garden might be bigger or perhaps smaller than mine. The size isn't all that important. What matters is that your garden is unique and will be exactly what you need to grow food for your family!

My mobile garden fits perfectly on my sunny deck.

GARDENING POWER

Did you know that throughout most of the history of the world, people have grown their own food? Even wealthy people who owned their own land hired others to work the land and grow food for them.

Yet, they usually knew a lot about how to best grow food. Most everyone knew how to garden.

Only very recently, in the last hundred years or so, have we depended on people we don't even know to provide us with food. We rely on grocery stores to feed us and trust there will be plenty of food in the stores at all times, even though we're unsure of how or where they get their food. We assume the farmers and ranchers across the world will take care of our need for nutrition. Yet we have no idea what kinds of methods they use to grow the food we eat. In other words, we put our health and very lives in the hands of people we don't know. We're dependent on them. If we learn to grow our own food, however, we become less dependent on people we don't know.

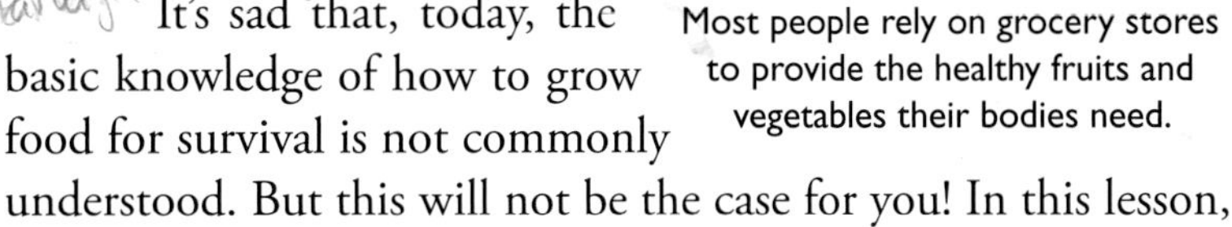

Most people rely on grocery stores to provide the healthy fruits and vegetables their bodies need.

It's sad that, today, the basic knowledge of how to grow food for survival is not commonly understood. But this will not be the case for you! In this lesson, you'll learn to grow your own food and will recover a lost art you can enjoy for the rest of your life!

This girl is experiencing the satisfaction of growing her own food to eat!

But that's not the only reason learning to garden is a great thing for you to do!

BETTER FOODS

Did you know that many of the fruits and vegetables you buy from the store are loaded with harmful chemicals? Most of the chemicals are pesticides the farmers use to keep bugs off the fruits and vegetables while they're growing. Some say that a single apple may have four different kinds of pesticides on it. When I was a child, I couldn't eat apples without getting a blister on my lips. I didn't realize it then, but it was the pesticides that were giving me the blister. When I ate an organic apple (an apple grown without the use of chemicals and pesticides), I didn't get a blister for the first time in my life! It was then that I realized how harmful pesticides are for us.

If you grow your own garden, you can do so without pesticides. In this lesson I'll teach you to grow an organic garden. You'll also learn to use natural methods of killing bugs that aren't harmful to humans.

In addition, your fruits and vegetables will be much more nutritious and taste a great deal better. Often, store-bought fruits and vegetables are picked by machines. Because these machines are so rough on the fruits and vegetables, they must be harvested or picked before they are fully ripe. When a fruit isn't quite ripe, it's firmer and able to withstand rough machine harvesting. However, when it's not allowed to ripen on the plant, the fruit doesn't receive all the nutrients it needs from the soil and air, and it doesn't taste as good. But that's not the only reason fruits and vegetables from the store don't taste as good as homegrown vegetables.

Underripe fruits and vegetables appear dull and lack both nutrients and taste.

These ripened homegrown tomatoes are more flavorful and nutritious than underripe store-bought tomatoes.

Grocery store foods aren't as flavorful because farmers don't use the same quality soil that you will use to garden. It would be too expensive for farmers to garden with rich soil. Instead, most use regular dirt and load it with manmade chemical fertilizers to feed the plants. These manmade nutrients rob the plants of the natural nutrients found in rich soils. This is why homegrown tomatoes taste sweet and juicy and have a strong tomato smell, while store-bought taste rather bland and barely have a smell at all.

When I go to the store to get tomatoes, I always choose those with the strongest smell. That way I know I'm getting produce that was able to ripen longer on the vine.

A final reason for growing your own food is you can save a lot of money on your grocery bill if you do it right. Store-bought fruits and vegetables not only taste bland, they are also quite expensive.

As you can see, growing a garden will be quite beneficial. I hope you'll enjoy the fruits of your labor as you embark on this journey!

Tell someone a few good reasons to grow their own food.

ACTIVITY 9.1

CREATE A GARDEN JOURNAL

You'll have some great successes as you grow your own food, but you'll also have some disappointments when things don't work out as planned. That's okay. In the end, when you finally see and taste the food you've grown, you'll be thankful for all the time and effort you invested in your garden! It's fun and useful to record your gardening experiences in a journal. Let's create a garden journal just like the nature journal you created in Lesson 1. In this journal you can record everything you did to create and maintain your garden, including what worked and what didn't work. This will help you become a very successful gardener!

ACTIVITY 9.2

PLAN YOUR GARDEN

In your garden journal, make a list of vegetables and fruits your family eats regularly. For example, if your family enjoys salsa, write down all the food items in salsa, such as tomatoes, onions, cilantro, jalapeno, tomatillos, serrano peppers, or onion.

Now go through your list and underline those items that would be ideal to grow in a garden. Ask your family what they would like to grow as well. This is the beginning of your garden plan.

My mobile garden is filled with our family's favorite fresh salsa ingredients: tomatillos, cilantro, and jalepenos!

TOOLS FOR GARDENING

There are a few things you'll want to have before you begin gardening. Some are just suggestions, but others are necessary for your garden's success. These tools can be purchased at most stores, but you can find everything you need and more at a plant nursery.

GARDENING GLOVES

Gloves are so important when it comes to gardening. They keep your hands clean and protected, making gardening easier and safer. Gardening gloves come in kids' sizes and can be purchased almost anywhere.

TROWEL

A **trowel** is a small shovel. It's an important gardening tool. You'll need a trowel to dig holes for planting and also to transplant plants that have outgrown their spaces.

WATERING CAN

You'll need to water your garden often. A watering can is helpful because water hoses release too much force. When you water, you're goal is to dampen the soil, not the plant. With a watering can, you have more control over where the water goes.

PRUNING SHEARS

As your plants grow, you'll need to prune them to ensure they yield the most fruits. Excess branches and shoots steal nutrients from the main stem and hinder fruits from forming and ripening.

EMPTY SPRAY BOTTLES

Once you begin growing luscious plants, you'll probably have to fight off the pests and diseases that want to feed off your plants. There are many natural ways to do this, but most require mixing solutions that will be sprayed on your plants. You'll want several empty spray bottles to mix solutions that will get rid of the diseases and pests that plague your garden.

PLANT MARKERS

Although you may think you'll remember what you planted where, it's often a good idea to have plant markers to label your plants just in case you forget. You can use popsicle sticks or buy plant markers from a nursery.

RAISED BED GARDEN

Although some people pick out a sunny plot of land, till the soil, and grow their garden in the ground, you're going to create a raised bed garden instead. The reason is a raised bed garden allows you to have total control over the soil your vegetables grow in. The soil we use for our plants is so very important, and we don't want anything to ruin our nutritious soil. Even if you only have a small patio or porch, you can build a raised bed garden for it.

A raised bed garden is also warmer, which means your plants have a longer period of time to grow and produce. You simply need to choose the sunniest location in your yard where you will place your raised bed garden.

LOCATION

Where you place your garden is very important. You'll want to choose a spot that receives around 10 hours of sunlight a day. Six hours is considered full sun, but vegetables need a bit more than that.

It's a good idea to put your raised bed in a spot that you walk by often. That way you won't forget to water it!

BUILDING YOUR BED

You can use any material to create the sides of your garden. People have used all kinds of supplies to make their beds: cinderblocks, railroad ties, all kinds of woods, and even plastic. You can choose whatever material works for you, or whatever you have on hand.

I've chosen to use regular pressure treated lumber for my garden, as it's inexpensive. Although some people might worry about chemicals in the lumber seeping into the soil, most scientists believe it's such a small amount that it's not a problem. Most people believe cedar is the best wood to use because it lasts longer.

Whatever you choose to use, make sure the sides are at least seven inches high because you'll need to have six inches of soil in your garden. That's because most vegetable plants need about six inches for their roots to grow.

WIDTH AND LENGTH OF GARDEN

I recommend you make your garden not much wider than three feet. That way, you can easily reach your hands across the garden from one side. But you can make it as long as you want. You'll need to figure out how much sunny space you have before deciding.

Because I only have one small area outdoors that receives enough sun to grow a vegetable garden, I'll be making a smaller bed of three by four feet. If you have a lot of space, you can create a longer bed with many more plants, or you can even create several raised beds!

ACTIVITY 9.3

BUILD YOUR RAISED BED

Are you ready to start your raised bed? Let's begin by building a box of some sort. Here are some examples of how I built mine. You can go to the course website for other ideas on how to build your raised bed garden. The key is to make sure the bed is raised between six to nine inches above the bottom of the bed. That way, your roots will have plenty of room to grow. When you're finished, continue reading to learn what soil you should use.

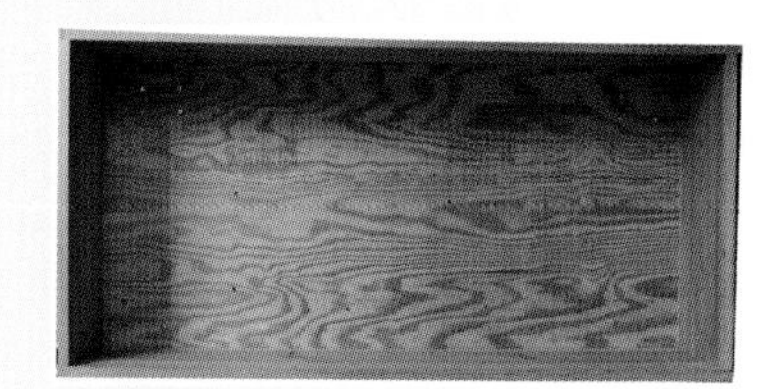

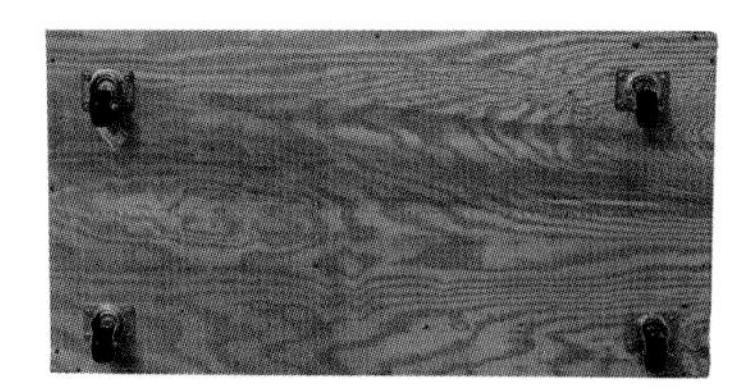

SOIL

Creating the best soil mixture to grow your plants is one of the most important things you must do. A good soil mixture will give your plants everything they need to survive and thrive. With the right mixture, you won't need to add fertilizers and other chemicals to your garden at all. This means you can grow your food naturally and organically!

The mixture you'll use is called the Pea-Ver-Comp mix. This name will help you remember the contents of the mixture because *Pea* refers to peat, *Comp* refers to compost, and *Ver* refers to vermiculite. With soil made up of equal parts of these three substances, your plants will be happy and healthy, having everything they need to thrive. Let's explore why each of these is important to your soil.

PEAT

Peat is a kind of dirt that has many nutrients in it. It's also especially good at trapping moisture. That means it will keep your garden from drying out too quickly and will also prevent too much water from killing your plants' roots. With a mixture of peat in your soil, you save your soil from drying out or getting too wet.

VERMICULITE

Vermiculite is a mixture of tiny mineral flakes that soak up a great deal of water. Because the mixture holds so much water, your plants have less chance of drying out in the hot sun. Vermiculite also provides phosphorus for the plants. Do you remember how phosphorus helps your plants? In addition to phosphorus, vermiculite contains special nutrients such as calcium, magnesium, and potassium, which are important nutrients your plants will need as they grow.

COMPOST

Composts act as fertilizers for organic gardens, keeping the plants growing strong.

As you already learned, compost is decomposed and decayed plant material that's super rich in the nutrients that feed plants. This is your soil's most important ingredient because it's your plants' actual food. By using compost instead of plain dirt or garden soil, you won't need to fertilize your garden with chemicals. The right mixture of different kinds of compost will give your plants all the fertilizer they need. This is the key to organic gardening and growing strong plants.

You'll need to use at least two different kinds of compost. This will give your plants a variety of nutrients. There are many different kinds of compost you can buy from the store—for example, mushroom compost (made of mushrooms), manure compost (made of horse, cow, or sheep droppings from the digested plants they consumed), vermin compost (made of earthworm droppings), yard waste compost (made of grass and leaves), or garden compost (made of plants). Even better than all of these is a garden compost that you make yourself as described in Lesson 7.

MIXING YOUR SOIL

When mixing the three ingredients above, you'll need to add equal amounts of each of these materials. To ensure the amounts are exactly equal, you'll measure them by inches.

As I mentioned, your garden should be six inches deep. You'll add two inches of peat, two inches of vermiculite, and two inches of compost. If you choose to have a deeper garden, simply divide up the number by three and add that much of each material to your garden!

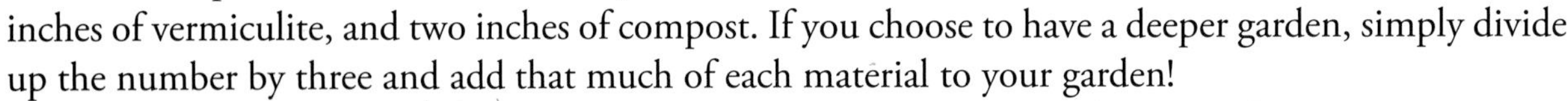

After measuring and adding equal parts, be sure to mix your soil ingredients well.

After you've added all three ingredients in the correct amounts, you'll mix them together really well using either your hands or your trowel. Voila! You have exactly what your plants need to grow and thrive!

When your plants have completed their growing season you'll want to remove them and add them to your compost pile. You'll only need to add more compost to your garden for the next growing season. The reason you only need to add compost is because the plants will likely have used all the nutrients in the compost. However, the vermiculite and peat will still be present and will not likely need to be replaced.

Explain to someone why the quality of soil you use is so important to your plants.

WHEN TO PLANT

Now that your garden is ready, it's time to plant! Deciding when to plant is as important as deciding what and where to plant.

Where do you live? Is it cold there much of the year? Is it warm? Is it mild? Is it hot all the time? Where you live is called your zone. When you plant your garden will depend on your zone. Look at the region map of the United States below. It's called the USDA Hardiness Zone Map. That's because if a plant can survive through the winter in that area, we say it's hardy in that zone. Each zone is given a number. When you buy plants, that number will often be listed on the plant's container or tag. This will tell you whether or not a plant will survive the winter in your area. If you live in zone 4, but the plant is only hardy in zones 7 through 11, you can only grow that plant during the warm growing season. If you live in the United States, take a minute right now to find your zone and its number on the map.

USDA HARDINESS MAP

Most states in America have a growing season from May to September. The Southern part of the United States has a longer growing season, from March or April to October or November. In South Florida, people can grow most plants year round, as can people in some areas of California and Arizona. Your growing season depends on when it's likely that freezing weather will occur. We call this freezing weather frost. The growing season is the time of year when the threat of a freeze for your area is over. Study the chart below to see when your growing season might be, and when you can begin planting your garden.

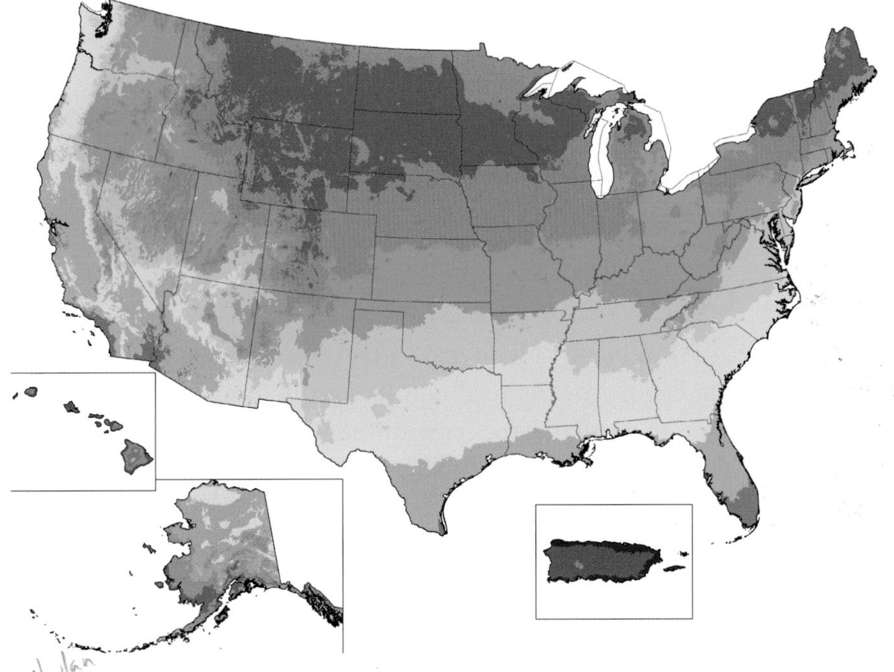

Average Annual Extreme Minimum Temperature 1976-2005

Temp (F)	Zone	Temp (C)
-60 to -50	1	-51.1 to -45.6
-50 to -40	2	-45.6 to -40
-40 to -30	3	-40 to -34.4
-30 to -20	4	-34.4 to -28.9
-20 to -10	5	-28.9 to -23.3
-10 to 0	6	-23.3 to -17.8
0 to 10	7	-17.8 to -12.2
10 to 20	8	-12.2 to -6.7
20 to 30	9	-6.7 to -1.1
30 to 40	10	-1.1 to 4.4
40 to 50	11	4.4 to 10
50 to 60	12	10 to 15.6
60 to 70		15.6 to 21.1

GROWING SEASONS BY ZONE

Zone 1 can only grow vegetables from June to July
Zone 2 growing season is May through August
Zone 3 and 4 growing season is May through September
Zone 5, 6 and 7 growing season is April through October
Zone 8 growing season is March through November
Zone 9 growing season is February through December
Zone 10 growing season is year round, except when there's a frost in late December or January.
Zone 11 growing season is all year round without a threat of frost.

WHAT TO PLANT WHEN

Knowing when to begin placing your plants in the ground is very important. One single freeze can destroy all your plants in one night; however, there are some plants that can handle a frost and will not die. You must know which of your fruits and vegetables can survive a frost and which cannot if you plan to begin your garden before the last frost. Go online to the almanac.com website that lists plants and when you can safely plant them in your specific area.

Explain to someone all that you've learned about growing seasons and zones.

ACTIVITY 9.4

JOURNAL YOUR PLAN

After you've researched what to plant when, write down your planting plan in your garden journal. When will you plant each plant you've decided to grow?

PLANTING SEEDLINGS

If you're a few months away from being able to grow plants outdoors and are anxious to get going, you can begin growing seedlings in your light hut. However, be sure to transfer them into small containers with the Pea-Ver-Comp mix after they've sprouted so they can grow into seedlings that are ready to plant when the time is right.

If you're ready to plant your garden now, I recommend purchasing seedlings from a nursery or online garden store. When planting a seedling, remove the lower leaves and set the plant in a hole made in your soil. Then cover the roots and a small part of your stem. Be sure to water your seedling well!

Seedlings like these can be planted and kept in your light hut until the outdoor growing season begins.

SPACING PLANTS

How far apart you space your plants is very important. If you plant too many plants in too small of a space, they will not produce well. That's because they will be competing with one another for sunlight, water, and nutrition. It's better to have one tomato plant producing ten red tomatoes than five tomato plants that can't ripen a single tomato.

Also, spacing your plants far enough apart will help the plants resist diseases. Diseases can develop when plants are placed too close together because there is little air circulation. Besides the fact that plants must have air, the high humidity of crowded plants encourages the growth of mildews, fungi, and bacteria that can kill your plants.

Proper spacing will protect your plants from disease.

To know how to space a plant, you'll need to know how big the plant can grow. Look at the information on the back of the seed packet or check the seedling information sheet for spacing instructions. If you don't have that information, go online to find web pages that tell you this information.

If you choose to grow vine plants, such as squash or pole beans, decide whether you want the vines to grow upward on a trellis or pole, or outward to trail along the ground. If you choose to grow them outward, plant them on the outside edges of your garden so they do not take up space inside your garden. Some plants, such as tomato plants, get very heavy with their fruit and must be stabilized with cages or poles so they don't fall over. There are many options for creating structures on which to grow and stabilize your vine plants and heavy fruit plants. Go to the course website to find links with different ideas for creating these structures.

Pumpkins grow well on heavy trellis structures like the one shown here.

TALL TO SMALL

If a plant requires a lot of sun but a tall plant is casting shade over that plant most of the day, it will not produce. So deciding where to place the plants in your garden based on how tall they grow is very important. Do some research to find out the potential size of your chosen plants. Do they grow into large bushes? How tall do they get? Do they need a lot of shade (like lettuce and cabbage), or a lot of sun (like tomatoes and peppers)?

Because of the way the sun rises in the east and sets in the west, it's important to place tall growing plants or upward growing vines on the north or west side of your garden. This way, the tall plant does not block the sun from shining on the other plants during the day.

ACTIVITY 9.5

MAP YOUR GARDEN

Get a compass (most smartphones have one) and discover where the north side of your yard and garden is. Draw a picture of this in your garden journal. Now create a map of where to place your plants in your garden based on how big and tall they're expected to grow.

Be sure to thoroughly soak your plants' roots in the cool of day.

WATERING YOUR GARDEN

Just like you need water to survive, your fruit and vegetable plants must have plenty of water each day to live. What happens if your soil has too much water? Well, just like you, your plants' roots can drown if they're not getting oxygen. Using the Pea-Ver-Comp mix prevents your plants from drowning because the vermiculite takes in the excess water and stores it for later use, allowing the plants plenty of oxygen to breathe. So with our special mixture it's not likely you'll drown your plants. However, you will need to make sure your plants don't dry out, especially in the hot summer months. The best way is to check the soil daily to make sure it's very moist. You can usually tell if your plants aren't getting enough water because the leaves begin to look droopy and sad. Keep an eye out for droopy leaves.

The general rule of thumb is that vegetable plants need an inch of water per week. You should thoroughly soak the soil each time you water the plants, rather than sprinkle a little water now and then. When watering, be sure to aim for the roots. The leaves don't need water. Also, when you water the portion of the plant that extends above the soil, you risk your plants developing diseases. Moist leaves can attract and grow certain diseases, and these diseases can spread through water droplets. It's best to water in the early morning or late evening. That's because if the hot sun is shining, a lot of the water will evaporate before ever reaching the roots.

ACTIVITY 9.6

MAKE AN IRRIGATION SYSTEM

One way to help with watering your garden is to create a homemade irrigation system using plastic bottles. This makes watering the roots easier. Let's create your own irrigation system now!

You will need:

- Empty plastic bottle with lid (sturdier energy drink bottles with a wide mouth work best)
- Drill or hammer and nail

You will do:

1. Drill or hammer five holes around the bottom edge of the bottle.
2. Bury the bottle in the soil of your garden in between your plants.
3. Fill the bottle to the top with water.
4. Put the lid on the bottle so that the water does not evaporate before it can water the plants.
5. Refill the bottle as needed.

MAINTAINING YOUR GARDEN

You've learned the importance of making sure your garden is planted correctly and well watered. But there are a few other things you should do to keep your garden healthy and growing throughout the entire season. Let's take a look!

PRUNING

If left to grow naturally, a plant will grow bigger and fatter, rather than use its energy to produce flowers, which, as you know, turn into the fruit we want to eat. The best way to encourage your plants to put their energy into the flowers and fruit is to remove the excess branches, stems, and leaves that pop up on the main stem. But don't go crazy pruning. Just prune a little here and there and your plants will thank you for it. Let's explore some techniques for pruning.

Snip Extra Stems

When the plant is young, snip off some of the shoots growing off the main stem. Pruning the stems greatly benefits plants that grow tall and wide, like tomatoes, vines, or squash. If a vine has too many stems growing in all different directions, not only is too much energy taken from the plant, it's harder to keep the vine upright and growing properly up the trellis or support rods. Also, having fewer stems on one plant will allow more airflow and sunlight to reach the plant's lower branches, flowers, and fruit. More sun, more air, better plants!

Some gardeners only allow their tomato plant to grow one single large stem. You can experiment to see how to prune your tomato plants to produce the most tomatoes. Be sure to include your results in your garden journal.

Removing suckers ensures the plant has all the nutrients it needs to produce healthy fruit.

Pinch Suckers

Suckers are the little leaves growing from the crook where two branches form a "Y." Suckers suck up important nutrients while they are growing, robbing the plant of what it needs to produce healthy fruit. You can simply pinch off suckers. Be careful not to damage the Y because that will damage the plant.

Pinch Shoots

Pinching off new **shoots** growing on the stems will allow your plant to focus its energy on the flowers and fruits instead of growing new branches. Shoots are extra stems growing off the main stalk. It's better to have fewer shoots so the plant can focus its energy on growing fruit.

Thin Flowers

Another way to increase the quality of your plant's fruit is to thin out the number of flowers you see on your plant. Too many flowers will take a lot of energy from the plant and the fruit won't ripen as fast. If you pinch a few flowers off the plant, it can focus on developing and ripening bigger, better fruit more quickly. The rule of thumb here is to pinch off a fourth of the flowers growing on the plant. So, if you have four flowers, pinch off one.

Thinning out flowers enables the plant to direct more energy toward growing and ripening its fruit.

It's important to remove poorly growing fruit to protect the plant and other plants in your garden.

Pinch Poorly Growing Parts

Regularly pinch off any deformed, diseased, or ruined fruit. Watch for insects or animals that may be feeding on your fruit and remove them, allowing the plant extra energy to put toward the healthy parts. Diseased fruit can spread the disease to other plants in your garden.

Clip Low-Hanging Leaves

Leaves touching the soil often become damaged and diseased. Clip them off in order to preserve the health of your plant.

Watch for low-hanging leaves and clip them off when you see them.

Top Before Frost

When you're about a month away from the end of your growing season and you have a bunch of unripened tomatoes and other fruit, clip off the tip of the main stem. This is called topping. It stops the plant from attempting to make more flowers so it can put its energy into ripening the fruit currently on the plant instead.

Before the first frost, trim back all the branches on the plant except for those that have fruit. Cut off the flowers and also the fruit that are too small to fully ripen before the season ends. This way your plant has a chance to give you fully ripe fruits before the first frost.

PEST PROBLEMS

Even though healthy plants can usually resist pests, you might still encounter some along the way. Here are some of the culprits and ideas for getting rid of those creepy critters intent on spoiling your garden.

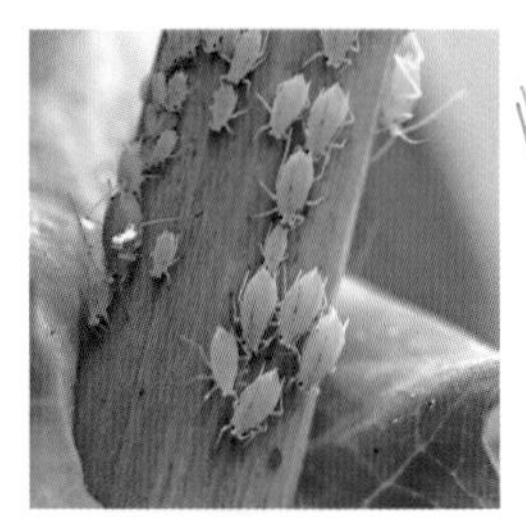

Aphids like these are a threat to your plants' health.

Aphids

You can identify **aphids** by their tiny, pear-shaped bodies, their long antennae, and the two pike-like projections on their back. They love the juice of leaves, fruits, and vegetables, as well as flowers. Aphids suck up the sugary sap a plant produces, causing the parts of the plants they invade to die.

To rid your plants of these pests, mix together one tablespoon of vegetable oil, five drops of Ivory soap, and a quart of water. Put the mixture in a spray bottle and spray the underside of the plant's leaves.

Mites

Mite colonies feed on the underside of leaves, causing them to turn yellow and fall off. These tiny creatures are usually reddish in color and look like little spiders. That's because they, like spiders, are arachnids.

To get rid of mites, mix together two tablespoons of hot pepper sauce, five drops of Ivory soap, and a quart of water. Put the hot sauce mixture in a spray bottle and spray the underside of the plant's leaves.

Keep an eye out for mites on the underside of leaves.

Caterpillars and beetles can be controlled with garden fabric.

Beetles and Caterpillars

The best way to handle beetles and caterpillars is to simply pick them off the plants as you notice them. However, if it's a widespread problem, cover the plants with garden fabric. Make sure you don't touch the caterpillar with bare hands because some can cause rashes or blisters.

DISEASE PROBLEMS

Even healthy plants get diseases sometimes; it comes with gardening. Don't get discouraged if you notice your plants' leaves getting spots or turning yellow. Figure out why and fight the problem. If you catch the disease early and treat it quickly, you'll nip the problem in the bud. Let's look at some of the more common diseases you might encounter.

This diseased pepper should be treated immediately.

Mildew

Mildew looks like white or gray powdery spots on the plant's leaves or stems. It's common on tomatoes, squash, asters, cucumbers, and lettuce.

Milk spray is a great treatment for mildews. The milk changes the pH on the leaves, making it a bad environment for the growth of mildew. To make milk spray, mix one part milk to nine parts water. Put the mixture in a spray bottle, and spray the plants thoroughly every three to four days.

This plant has been overtaken by mildew.

Baking soda is another helpful treatment for mildew and fungal problems on your plants. Simply mix two tablespoons baking soda into a gallon of water. Put the mixture in a spray bottle, and spray the affected areas every few days.

Fungus

Rust is a fungal infection that looks like a rusty brown powder or hairy substance on the plant's leaves. In the early stages, it looks white, with slightly raised spots on the undersides of the leaves and on the stems.

Leaf spots are either fungal or bacterial infections that cause brown or black spots on the leaves, sometimes with a yellow border. As the number of spots grows, the leaves die. And we don't want that, do we?

The spots on this leaf reveal a fungal or bacterial infection.

One solution for leaf spots is apple cider vinegar. Pour one tablespoon of vinegar in one gallon of water and put the mixture in a spray bottle. Spray the solution on the affected leaves.

Bacteria

Bacteria are always present on plants. They usually don't cause problems unless the plants are overcrowded. Overcrowding, as you may remember, causes high humidity, poor lighting, and poor air circulation. Keep your plants spaced nicely to avoid bacteria growing on your plants.

If you do see bacteria, hydrogen peroxide is a great cure for both bacterial and fungal problems. If you have a solution of three percent hydrogen peroxide, it can be sprayed directly on the leaves to treat and prevent these problems. Don't use it on seedlings, however. It's only safe for mature plants.

Here are some other important tips to keep your garden free of pestd and disease:

- Clean your tools with disinfectant after using them, especially if you you've used them on diseased plants.
- Always use clean containers and soil.
- Don't combine regular dirt from your garden with your soil. You don't know what diseases that dirt might contain. Always use fresh soil that you have mixed yourself.
- Always allow plenty of space between your plants. If they get crowded, thin them out by transplanting a few in other places. An overcrowded space is the perfect environment for diseases to grow.
- Use your observation skills to study your leaves and stems carefully. Take note of anything that looks suspicious and take action immediately before the problem gets out of control.

These are just a few of the numerous natural solutions you can use to fight diseases and pests on your plants. Go to the course website for links to other methods for treating your garden.

ANIMAL PROBLEMS

One problem I had with my garden was pesky visitors: an opossum during the night and sneaky squirrels during the day. Many people have problems with deer. It's been said that deer don't like the hot sauce mixture I mentioned in the mite section above. If you find furry animals stealing the fruits of your labor, use garden fabric to cover the plants. However, be aware that pollinators can't reach your plants if they're covered. So you may want to leave them uncovered for part of the day. Garden fabric is sometimes called floating row covers. You can place the fabric directly on the plants, or you can use hoops to cover them as I've done.

Hungry deer are common to backyard vegetable gardens.

FALL AND WINTER IDEAS

When your spring and summer plants are about to end their production, it's time to begin thinking about your fall season. There are many plants that can be started in the fall and will even stand a frost or two. You might try beets, broccoli, cabbage, carrots, collards, lettuce, kale, onions, radishes, spinach, or turnips.

A garden cover is a simple solution for extending your growing season.

When the time is right, remove your spring plants and throw them in your compost pile. Then, add more compost to your Pea-Ver-Comp mix and plant your fall garden.

To lengthen your growing season, warm up your garden by covering it with the garden fabric I mentioned above. This special fabric allows sunlight and water to enter your garden and also increases the temperature by holding in the heat. You can drape it right over the plants or use hoops to hold it above the plants as explained in the pest section.

You're now well on your way to becoming a true gardener! I hope you enjoy your gardening project and are able to produce a lot of fruits and vegetables for your family. Enjoy!

WHAT DO YOU REMEMBER?

What are some good reasons to grow your own food? Which side of the garden should you plant tall-growing plants? What zone are you in? What is your best growing season? Name some tips for watering your garden. Why is pruning important? What are some solutions for getting rid of pests in your garden? Name a solution for getting rid of diseases on your plants.

ACTIVITY 9.7

DRAW YOUR GARDEN

In your Botany Notebooking Journal, after recording the facts you learned in this lesson, draw a diagram of the raised bed garden you are planning. Show where you will place each plant. Include a small compass rose in the corner for direction.

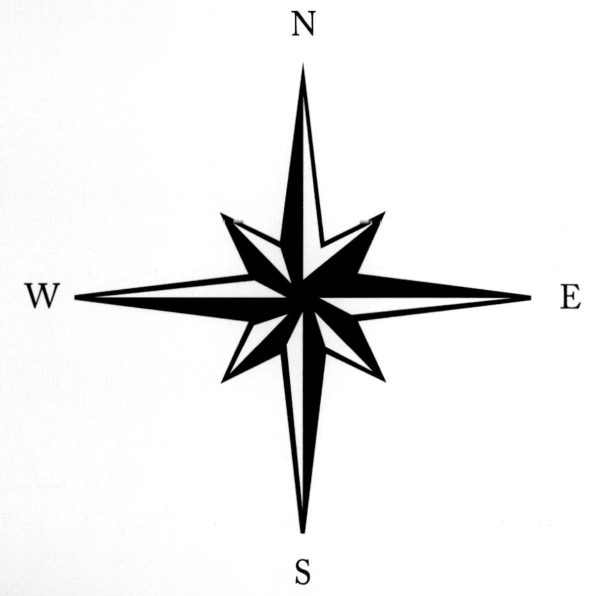

LESSON 10

TREES

digging deeper

Christ redeemed us from the curse of the law by becoming a curse for us, because it is written, "Cursed is everyone who is hung on a tree." Galations 3:13

What do you think it means that Christ redeemed us from the curse by becoming a curse for us? It means that because of sin—the wrong things we have done, the bad things we have thought, and the ways we have failed to do right—we are cursed or guilty before God. You see, all sin, even the tiniest sin, keeps us from entering God's perfect heaven. But Jesus did something amazing. He took the punishment for our sin. He traded his sinless perfection for our sinful imperfection. He made us clean, washing away our sins. At the same time, He gave us the credit for His perfect obedience to God's law. Jesus did this when He hung on the cross so we could go to heaven. And that cross was made out of a tree. Whenever you see a tree, remember to thank Jesus for dying on the cross and freeing you from the curse so that you can go to heaven and live forever with Him.

TREES ARE SPECIAL

Do you remember the story of Zacchaeus, the short man who climbed a sycamore tree to get a glimpse of Jesus passing by (Luke 19:1–9)? Jesus stopped, called Zacchaeus' name, and told him He would be eating at his house that day. Would it surprise you to learn that the sycamore tree Zacchaeus climbed in Jericho is not the same sycamore tree we have here in America? That's because common names for plants are different all over the world which is why we use Latin to describe living things. The tree that Zacchaeus climbed was probably the sycamore fig tree, a *Ficus sycomorus*. The Latin name for the sycamore tree we have in America is *Platanus occidentalis*. At the end of this lesson, you'll find out the scientific and common names of all the trees in your area. When I researched the trees in my area, I discovered I have two species of oak trees, a white oak and a red oak. I can tell the difference between the two based on the slight variations of their leaves. I also found that the chestnut tree in my yard is actually a hickory tree. This is a really fun exercise and will help you become an even more educated botanist!

Have you ever climbed a tree like Zacchaeus? When I was a child, I spent a good deal of time in the live oak trees that surrounded my house. There were so many trees, and my brothers and I each claimed one as our very own. I would sometimes climb my tree and snuggle down into a large recess between the branches to read a good book.

My own children have a special tree that they often climbed to the very top of when they were growing up. A rope swing tied to that tree provided hours of fun and many memories for my children. Trees are very special. Do you have a special tree in your yard? If so, try to draw it in your nature journal.

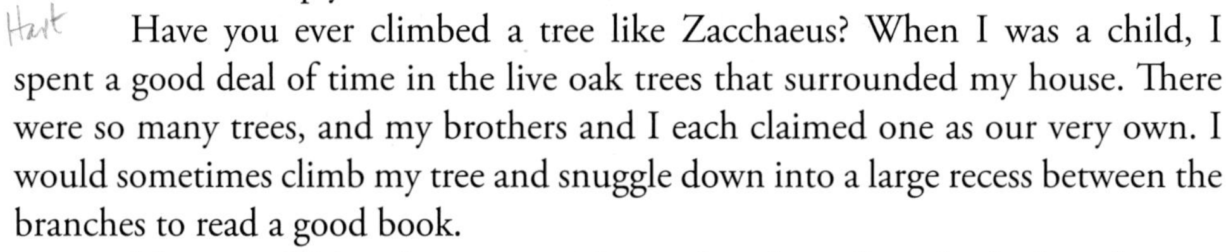

Trees are some of God's most special creations, providing enjoyment in many different ways.

TREES ARE IMPORTANT

Do you realize that the oldest living things on Earth are trees? In fact, there are trees in California that were alive when Jesus was on Earth. Believe it or not, they were alive long before then! In this lesson you'll learn all about trees and their importance in God's amazing creation.

God really gave mankind and the animals a special gift when He made trees. Trees provide the world with so much more than you might realize. They provide shelter, shade, beauty, food, and healthy air for humans and animals. As you already learned, their roots even help keep the entire structure of Earth's landscape stable.

Look around your home and try to count all the things that come from trees. Do you have windows in your home? The frames around those windows are probably made of wood from trees. Do you have wood on the outside of your house? Did you know that if you tore down all the walls in most houses, you would find that the entire frame of the house is made of wood? That means it comes from trees. What about your furniture? Is any of it made of wood? If so, it comes from trees. Even the book you're reading right now is made out of trees because we make paper from trees! We are, indeed, very dependent upon trees. As you remember, it was a tree that was made into the cross upon which Jesus hung to bear the sin of all who believe in Him!

ACTIVITY 10.1

IDENTIFY THINGS MADE FROM TREES

Look around your house and try to identify things that are made from trees. In your Botany Notebooking Journal, make a list of these things.

ANIMAL SHELTER

Trees are home to all sorts of animals: insects, birds, squirrels, lizards, and frogs, to name a few. Without trees, these animals would have little protection from predators (other animals that want to eat them). The tree is their sanctuary or home and, in some cases, their food supply. Many animals and insects eat bark, leaves, twigs, and fruit. If you look underground where the tree roots are, you'll find tunnels built by snakes, rabbits, moles, and many other kinds of burrowing animals. Some animals are even dependent on one single species of tree to survive. For example, the koala only eats the leaves from the *eucalyptus* tree in Australia. And the panda bear is dependent on the leaves and stem of bamboo, which is technically a kind of grass. However it can grow into a forest that looks an awful lot like trees.

The eucalyptus tree provides not only shelter but also food for the koalas.

Of course, trees aren't just important for animals. They're also vital for the health of humans. Because they perform a lot of photosynthesis, trees provide us with oxygen to breathe. Scientists have discovered it takes only two very mature trees to provide a single person with all the oxygen he needs for a lifetime.

Remember, during photosynthesis, trees consume chemicals in the air that make the air impure, like carbon dioxide. Scientists have estimated that in a single year, one acre of trees uses the amount of carbon dioxide a car produces after driving 26,000 miles. That's about twice the average number of miles a car drives in a single year. So a single acre of trees can clean up the emissions of two cars. If we want to clean up the air, we should plant more trees!

We can thank God for creating trees to help clean up harmful chemicals in the air.

Of the many reasons God created trees, we can see that having oxygen to breathe was an important one. You need to realize that while plants (including trees) provide us with a lot of oxygen, they aren't the only living things in creation that make this important chemical for us. In fact, as you learned in the photosynthesis chapter, more than half of the oxygen we breathe is produced by microscopic creatures living in the water performing photosynthesis. So even though trees and other plants do give us oxygen, it's important to remember there are other living things that give us oxygen as well.

Do you remember that plants help clean the air by removing pollutants? Well trees do a lot of that work as well. A botanist with the U.S. Department of Agriculture Forest Service studied several major cities to see how much pollution the trees in those cities removed from the local air. What he found might surprise you. The trees in Denver, Colorado, for example, remove just over a million pounds of pollutants each year! That's just one city. The more trees a city has, the cleaner that city's air! Trees clean the air by trapping and removing dust, ash, pollen, and smoke.

The trees surrounding Denver, Colorado help keep the air clean and healthy for people to breathe.

Trees are also important in protecting the Earth from erosion. You learned in Lesson 7 that the roots of plants help hold the ground in place. Because trees are

You can see how these trees' roots are helping to hold the ground above in place.

This tree's large roots anchor it securely to the ground.

so big and have so many roots, trees do a lot of that work. Without them, much of the Earth would be washed away into the oceans, lakes, and rivers during heavy rains and storms. Roots anchor trees to the ground, but they also anchor the ground, reducing how much it is reshaped by wind and rain. Trees are a wonderful addition to God's creation, aren't they?

Harleigh

MORE TREE FACTS

There are many other ways trees add value to the world. For example, researchers have discovered that areas with trees experience lower crime rates. Having more trees in your yard also increases your home's property value (how much the house costs). Students with trees surrounding their school have higher test scores. Because of transpiration and the shade they provide, trees can cool the temperatures quite a bit, saving much valuable energy. In the winter, trees can even bring protection from the wind. Do you now understand the extreme importance of trees?

Explain to someone why trees are so very valuable.

ACTIVITY 10.2

PLANT SOME TREES

Planting more trees will help the environment a great deal. You can plant a few trees by yourself in your yard by joining the Arbor Day Foundation. For a small fee, they'll send 10 trees that you can plant in your area. It would be even more fun to organize a group of people to plant trees in areas of your community where they're needed. If you're feeling really ambitious and want to help those suffering from tree loss, you can take a mission trip to an area devastated by floods or storms and plant some trees. Go to the Arbor Day Foundation website to learn how to organize a tree-planting project or how to join a tree-planting mission for devastated communities.

Harlan

SEED MAKING

Most trees are angiosperms. Do you remember what an angiosperm is? It's a plant that produces flowers. Many angiosperm trees produce what we call **imperfect flowers**. Do you remember what imperfect flowers are? We learned about them in Lesson 4. They have either male parts (stamens) or female parts (carpels), but unlike most flowers, they do not have both. Trees that produce imperfect flowers will produce certain flowers that have only the male parts (carrying the pollen) and also produce some flowers that have the female parts (carrying the eggs). As you know, the female flower will develop into

The abundance of acorns produced by this oak tree will fall to the ground as food for the squirrels.

a fruit (acorn, chestnut, etc.) once it has been pollinated.

Are there any trees in your neighborhood that produce acorns? If so, they are oak trees. Did you know that an oak tree cannot reproduce until it's at least 20 years old? Some wait until they're 50 years old before they produce their first acorns! Once an oak tree starts to produce acorns, it usually does so in a sporadic way. Unlike most angiosperms, which produce seeds regularly every season, an oak tree will have one year that it produces an enormous number of acorns. During such a year, the acorns simply cover the tree, and they fall from every limb. When this happens, they litter the ground with an overabundance of acorns. This is called **masting.**

Squirrels try to find masting oak trees because a single oak tree can produce as many as 2,000 acorns in one year! Since squirrels love acorns, they feast when they find a masting oak tree. For several years after that, however, the oak tree will produce only a few acorns, if any. Most oak trees go through masting about once every five years. The chances of one acorn actually growing into an oak tree are very slim—less than 1 in 10,000. That means that for every 10,000 acorns, only one will become a tree! This is because 9,999 out of 10,000 will be eaten or used by animals in some other way. For example, an insect called the **acorn weevil** eats acorns but also deposits its eggs inside them. The baby weevil, called a weevil larva, will grow up inside the acorn, eating its contents. If you find an acorn, chances are very good there will be a little weevil larva living inside it. In fact, you may find an acorn shell that's completely hollow and has a tiny hole in it. That acorn was probably home to a weevil larva that ate the inside of the acorn then burrowed out of it.

This weevil has found an acorn to feast on and will most likely deposit its eggs inside.

TREE GROWTH

As you may remember, once an acorn (or any other seed) begins to grow into a plant, it's called a **seedling**. When a tree seed is developing into a tree, it's considered a seedling until it grows to be a few feet high. At that point, it's usually called a **sapling**. There may be some saplings growing in your neighborhood. If nothing harms them, they'll one day grow to be mature trees.

Though only a few feet tall, this sapling is on its way to becoming a mature tree.

A tree grows in three different ways: the roots grow *longer* each day in search of water and nutrients. The tree grows *taller*, adding more cells to the tops of its branches as it reaches higher and higher. And the tree grows *wider*, producing a new ring of growth every year. Interestingly, botanists think there is a limit to how tall any tree can grow. Because it gets harder and harder for a tree to get water to its topmost leaves as it grows taller, most botanists think that trees can never grow taller than about 420 feet. The tallest living tree in the world is a redwood that is 369 feet high. It lives in the Humboldt Redwoods State

Park in California. Even if trees have a limit to how high they can grow, there is no limit to their width. Since a tree adds a new ring of growth to its stems each year, the longer it lives, the wider its trunk and other stems become.

These Humboldt Redwood trees are among the tallest in the world.

You can actually see how much height a tree has added each year by studying one of its branches. There are marks along the twig of a tree branch that tell us exactly how tall a tree grew in each previous year. Farther down the twig, the branch becomes harder and more woody. This is where thick bark is beginning to form. At first, the bark is smooth and sleek, but with age, it thickens, and the surface dries and cracks. Twigs don't have thick bark like the trunk. This makes it easy to tell how much growth took place in a year.

At the tip of each twig, you'll find a bud, called a **terminal bud**. In spring, these buds sprout forth a group of petioles and leaves. The branch adds cells to the twig all summer, making it grow longer. At the end of the growing season, a new terminal bud develops at the tip of the stem. On every twig, there's a circular scar where last winter's terminal bud was. You can find out how much the twig grew in one year by measuring the distance between the new terminal bud and the scar of last winter's terminal bud.

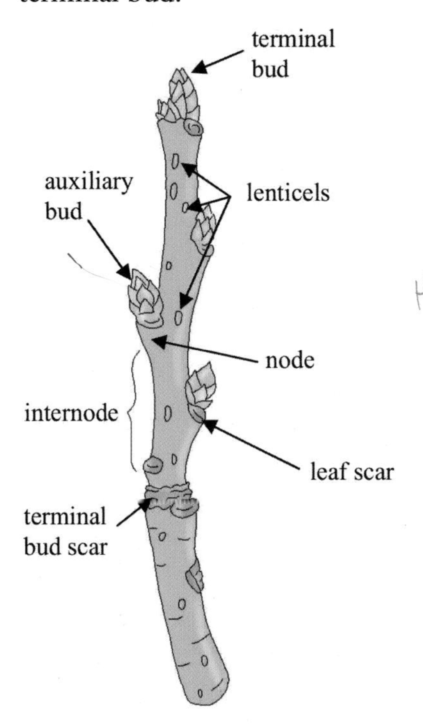

TWIG ANATOMY

Every twig has several features. Let's look closely at a single twig. If you can get one from your yard, do so now. It's best to find one in the fall or early spring before it's covered in leaves. As I explain each feature, look at your twig to see if you can locate what I'm describing.

As I already told you, the terminal bud is the bud at the very tip of the twig. There's a circular scar that surrounds the entire twig called the terminal bud scar, and it represents where the terminal bud was the previous year. **Nodes** are places on a twig where other buds are located. Petioles, leaves, and eventually branches will grow from these buds. They are called **auxiliary** (awg zil' uh ree) **buds** because they are not terminal buds located at the tip of the twig. Below the auxiliary buds are leaf scars where leaves once were. The spaces between the nodes are called the **internodes**.

There are also small nicks scattered throughout each twig. They sometimes look like little blisters. These are actually tiny pores where the tree takes in oxygen! Although a tree's leaves take in carbon dioxide and produce oxygen through photosynthesis, its cells actually need oxygen to survive, just like you and I do. These pores, called **lenticels** (len' tih selz), allow the twig to take in oxygen so that its cells can stay alive. I bet you didn't know a tree branch could breathe!

ACTIVITY 10.3
MEASURE TWIG GROWTH

To measure a twig's growth, you can use the twig you already have, but I recommend that you can go outside and find a new twig that's still attached to the tree (you don't need to remove it). Again, the best time to do this is during the fall or winter. It's possible to measure a twig's growth in the summer or spring, but the twig will be packed with leaves, making it harder to find all the features of the twig.

You will need:

- Ruler
- Colored pencils
- Twig on a tree

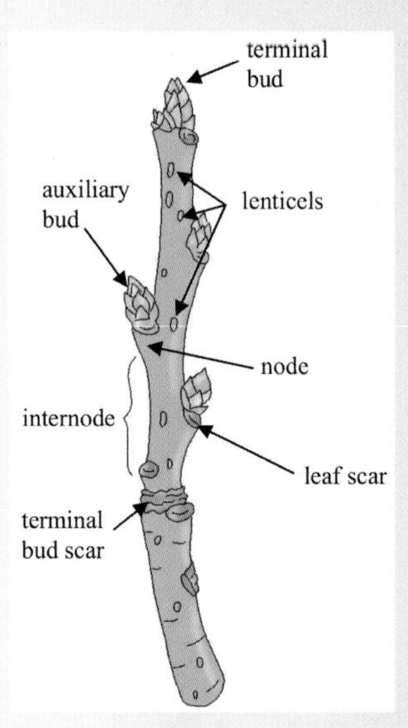

You will do:

1. Study the twig and try to find examples of all of the features pointed out in the drawing above.
2. Measure the distance between the terminal bud and the first terminal bud scar. That's the distance the tree grew during the previous spring and summer.
3. Moving down the twig, measure the distances between successive terminal bud scars. Which year did the twig grow the most? Which year did it grow the least?
4. Make a drawing. Be sure to label all of the structures you identified.

ACTIVITY 10.4
ESTIMATE THE HEIGHT OF A TREE

Have you ever wanted to know how tall a tree is? You might think there's no way to find out the height of a tree because you don't have a measuring tape long enough to measure it. Even if you did, how would you climb to the top of the tree to use it? You can get a very good idea of how tall a tree is with just a ruler and the help of a friend whose height you know.

You will need:
- Someone to help you
- Measuring tape or yardstick to measure the height of your helper
- 12-inch ruler
- Tall tree

You will do:
1. Use the tape measure to find out the height of your friend.
2. Take your friend outside and find a tall tree.
3. Have your friend stand right next to the tree trunk.
4. Walk about 15 feet away from the tree.
5. Hold the ruler at arm's length between you and the tree. Hold it so that the tree looks like it's right next to the ruler.
6. Pull the ruler back toward your face until the ruler looks like it's as tall as the tree. If the ruler appears smaller than the tree no matter how close to your face, walk farther away from the tree and try again. If the ruler is taller than the tree even when you're holding it out at arm's length, walk closer to the tree.
7. Once you've gotten the ruler to the point where it looks like it's as tall as the tree, hold it right at that position and turn your body (without moving your arm) so that your helper looks like he's right next to the ruler.
8. Use the numbers on the ruler to measure how tall your helper appears to be.
9. Take the measurement from step #8 and divide by the number 12. For example, if my helper appears to be 2 inches tall based on the ruler, I would divide 12 by 2 to get 6.
10. Take the answer you got in step #9 and multiply it by the actual height of your helper, which you measured in step #1. If my helper is 5½ feet tall, for example, I would take my answer from step #9 (which was 6) and multiply it by 5½ feet to get 33 feet. That's the height of the tree!

This activity makes use of a math concept called proportionality. By holding the ruler close to your eye, you made the ruler appear as big as the tree. When you used that same ruler at that same position to measure something with a known height (your helper), you were able to determine the height of the tree relative to the height of your helper. In the examples that I gave you, my helper appeared to be two inches tall using the ruler. Since the tree appeared to be the same height as the ruler, I knew that the tree was six times taller than my helper. Proportionally, then, the height of the tree was equal to six times the height of my helper.

GROWING OUTWARD

Start

Do you remember studying how liquids move through the stems of plants? Well the tree trunk is the stem of the tree. You might think that the liquid moves deep inside the tree trunk, but that's actually not the case. Almost all the activity happens just under its bark, close to the surface. There you find the phloem, and right under the phloem you find the vascular cambium. Right under that, you find the xylem. Do you remember what xylem and phloem are? The xylem suck the water up from the roots

Similar to your skin, a tree's bark is its protective covering.

and send it to the leaves, while the phloem cause the sugary food to flow down to be used and stored by the rest of the tree. Do you remember what the vascular cambium is? It's the layer of cells that makes the xylem and phloem, causing the tree's trunk to get wider each year. All this is happening right underneath the bark!

Because of all this important activity happening close to the surface, the bark of a tree is quite important. It's the shield of protection for the xylem, phloem, and vascular cambium. If you peel off the bark, you'll likely peel off a bit of the phloem as well, robbing the tree of some of its most important cells. What do you think would happen if you pulled all the bark off a tree? In some ways, it would be like peeling off all your skin! If someone peeled off your skin, you would lose water and nutrients, and you would certainly get diseases because your skin protects the inside of your body. In the same way, a tree's bark is its protection! Some trees shed their bark and replace it with new bark each year. If a tree has loose bark, it might be getting ready to shed it.

This tree is getting ready to shed its bark.

TREE TRUNKS

So what happens when a tree keeps adding layers and layers, getting thicker and thicker as it grows each year? Well, very deep inside the trunk of the tree is the oldest part of the tree. It's the heart of the tree. We call the wood on the very inside of the tree **heartwood**. In mature trees, the heartwood doesn't have any activity going on inside it. We say it's dead. It's wood that was once teeming with xylem and phloem. However, as layers of xylem and phloem were added each year, the old xylem got plugged up and died. When that happened, it became heartwood. Even though it no longer carries water for the tree, it still does an important job. The heartwood helps a tree survive strong winds by giving the tree strength. Sometimes, termites or wood ants eat away at the heartwood, making the tree mostly hollow on the inside. Trees without heartwood can continue to grow for hundreds of years, however, because the outer layers are the living parts of the tree. Of course, such a tree can topple over more easily because it doesn't have thick heartwood in the center to support it.

TRUNK LAYERS

A tree trunk can be separated into five layers. Each of these layers has a different job. Look at the image to your right as you read about what each layer does.

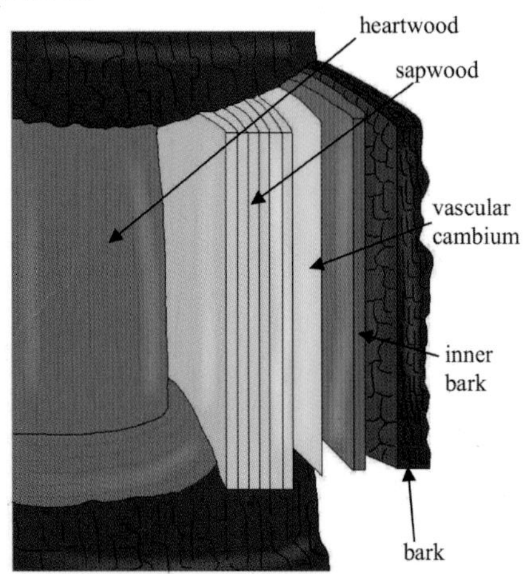

- The heartwood is the dead wood in the very center of the tree trunk. It gives the trunk strength.
- The **sapwood** surrounds the heartwood and is made of xylem that transports water up from the roots to the leaves. It also stores the excess food (starch) that the leaves make during the

summer. In the early spring, the tree starts converting that starch into sugar, and the sugar starts to run in the sapwood. The sugar that's flowing in the sugar maple tree's sapwood is the maple syrup we enjoy on our pancakes!

- The **vascular cambium** is a thin layer of cells that produces new xylem and phloem. It causes the tree trunk to get wider every year.
- The **inner bark** is made up of the phloem. It transports food from the leaves downward to the rest of the tree.
- The **bark** is the outer layer of the tree. It provides protection for the living parts of the tree. The outer bark is often called the **cork** of the tree.

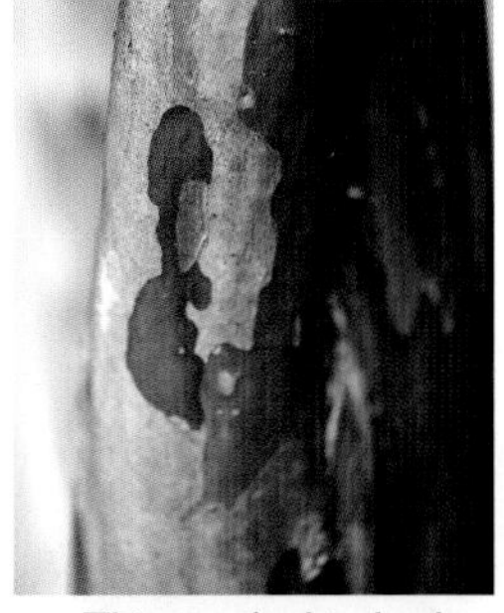

This tree's thin bark peels in layers.

TREE BARK PATTERNS

Every tree has its own unique bark. Scientists use the word cork to describe the bark of the tree. The patterns in the bark can be used to identify what kind of tree it is. Bark can be smooth or bumpy or spiny. It can be multicolored. It can also have long ridges or fissures in it. Birch trees have paper-thin bark that was once used by Native Americans to make canoes. The bark of the sequoia tree in California can be more than a foot thick with deep ridges!

The bark of young trees is smoother than that of old trees. As the tree grows, its bark can either get thicker and thicker or, on some trees, shed off as new bark replaces it. Because each tree's cork is so unique, people can identify most trees simply by studying their bark.

For thousands of years, people have used bark for many purposes. Bark is used to make canoes, rope, cinnamon, and aspirin, among other things.

Explain to someone all that you have learned so far in this section.

ACTIVITY 10.5
MAKE A BARK RUBBING

In this activity, you're going to make bark rubbings for your Botany Notebooking Journal. As I told you earlier, each tree has its own special pattern of bark, and you'll be able to copy this pattern by creating a bark rubbing. This will help you identify the tree. If you're making a field guide (described in an upcoming activity), you should include a bark rubbing for each tree you're including in your guide.

You will need:

- Crayon with all the paper removed
- Plain white paper
- Tacks (optional)

You will do:

1. Lay the paper across a tree trunk.
2. If you have tacks, use them to secure the paper to the tree. You can also just hold the paper with your hands, but tacks might make the job easier.
3. Holding the crayon sideways, rub it over the paper. As you do this, you'll see the crayon making marks on the paper in the pattern of the tree's bark. You just made a bark rubbing!
4. Put the bark rubbing in your Botany Notebooking Journal and note the kind of tree from which it came.

THIRSTY TREES

As you might have guessed, trees transport enormous quantities of water from their roots to their leaves. You'll be surprised to learn that the water can move at speeds of up to 25 feet per hour!

It's not unusual to see ferns growing on trees in the rainforest.

So what's the water used for? Some of it's used for photosynthesis. Do you realize that a tree uses less than 2 percent of the water taken in from the roots for photosynthesis? Do you remember what happens with the excess water? It transpires. That means it moves out of the leaf and into the air as water vapor. More than 90 percent of the water the tree takes in is transpired into the air! One large tree can capture and filter up to 36,500 gallons of water per year. That's the reason rainforests are so moist and humid. In fact, many plants and animals depend on this moisture to grow and survive. This helps us understand why ferns and orchids easily grow on the trunks of trees in these places.

Not all trees transpire that much, however. Some trees, called conifers, don't transpire much at all. You'll learn more about conifers in the next lesson.

Wow! You've learned a lot about trees! What was the most interesting thing you learned? Let's see if you can answer some questions before recording in your Botany Notebooking Journal all that you discovered today. After that, we'll begin our tree identification project.

WHAT DO YOU REMEMBER?

Name some important uses for trees. How can you tell how much a tree has grown by its branch? Explain the anatomy of a twig. Be sure to include terminal buds, lenticels, nodes, internodes, and auxiliary buds in your explanation. Describe the layers, names, and functions of a tree's trunk.

ACTIVITY 10.6

DIAGRAM A TREE'S LAYERS

After recording in your Botany Notebooking Journal what you learned in this lesson, diagram the layers of a tree's trunk. You can use the drawing earlier in the lesson as a guide. Be sure to label each layer and describe what it does for the tree.

ACTIVITY 10.7

MAKE A TREE FIELD GUIDE

You're going to create your own field guide for the trees in your area. God filled our world with thousands of varieties of trees, but there are really only two basic kinds of trees: angiosperms and gymnosperms. Gymnosperm trees make their seeds in open vessels like pinecones, and angiosperm trees make their seeds in closed vessels we call fruits. Gymnosperm trees almost always have needles (like pine trees) or scaled leaves (like juniper trees), while angiosperm trees have broad or flat leaves (like oak trees).

For this project, I want you to identify the trees in your yard or neighborhood using a field guide. A field guide often divides the trees into groups by the shapes of their leaves. If it's spring or summer, you can use leaves and fruits to identify the tree. Notice the tree's shape by looking at the shape of the trunk as well as the crown (top) of the tree. Is the crown shaped like an "A," a circle, or another shape? Most field guides show an outline or silhouette of the tree. This helps you to identify what kind of tree it is.

If you're in a cold region or it's the middle of winter, you'll want a field guide that identifies trees by their bark patterns, such as *Tree Bark: A Color Guide, Trees: North American Trees Identified by Leaf, Bark & Seed* (part of the Fandex Family Field Guides series), or *A Field Guide to Trees and Shrubs* (part of the Peterson Field Guide series).

Now it's time to create your own field guide! Use the same technique you used to create your nature journal in Lesson 1, or if you are adept at using the computer, create your field guide there and print it.

As you explore and identify a new tree, make a page for it in your field guide. Photograph or sketch the tree, taking note of its shape and size. Examine its leaves. Identify the shape, margin, venation, and texture. Make a special photo or drawing of its leaf. Does it have flowers or fruit? Describe them. Add an illustration of the flower or fruit. Discover whether the tree is deciduous (loses its leaves in the winter) or evergreen (stays green throughout the winter). Include in your field guide any additional interesting facts you learn about that tree.

LESSON 11

GYMNOSPERMS

digging deeper

The trees of the LORD flourish, the cedars of Lebanon that he planted. There the birds make their nests; storks make their homes in the pine trees.
Psalm 104:16–17

God cares for everything in creation. He provides homes for the animals by keeping the trees well watered and growing. How much more will He also provide for you? Put your trust in Him to always take care of your needs.

UNCOVERED SEEDS

So far in our study of plants, we've discussed some things that are common to all vascular plants, but mostly we've talked about angiosperms. Now it's time to move on and talk about the other type of seed-making plants in creation—gymnosperms!

A pinecone's seeds are found on the open slats of the cone.

While angiosperms create seeds inside a container, gymnosperms do something different. *Gymno* means uncovered, and as I told you already, *sperm* is another word for seed. This tells us that **gymnosperm** means uncovered seed. As you might imagine, gymnosperms produce uncovered seeds. They aren't completely without a shield, though. Let me explain: have you ever found a pinecone on the ground while exploring outside? Pinecones are grown on a certain kind of gymnosperm plant called a conifer. The word *conifer* means cone-bearer. The pinecone is what holds the seeds of the conifer plant. They are nestled on the slats of the open cone, however, since the cones are open to the air, the seeds are considered uncovered.

CONIFERS

Conifers are very special because they're the largest and oldest trees found on Earth. In fact, some conifers living today were seedlings when the earliest civilizations began emerging after Noah's flood. Many of the giant sequoias in the Pacific Northwest of the United States are more than 3,000 years old!

Although conifers grow in every climate, almost all the trees you find in cold regions are conifers. In fact, if you look at a globe, cone-making plants are the main trees you find from Canada across to Siberia and all the way to Norway. They practically circle the top of the globe!

Conifers are beautiful, cone-bearing trees.

The giant sequoia tree I mentioned earlier is

As you can see, these giant sequoias have very wide trunks and thick bark.

pretty interesting. These trees reside in the Pacific Northwest part of the United States. Not only do they grow to be really old; they grow to be enormously tall and wide as well. Some of these trees reach more than 300 feet into the sky. As I said in the last lesson, the bark of this tree alone can be more than a foot thick!

Giant sequoias are the largest members of a group of trees called redwood trees. They're named redwoods because they are somewhat red in color. In fact, the first Spanish explorer to see redwoods called them *palo colorado*, which means red tree. When the early settlers cut these mammoth trees down, they used the tree stump as a dance floor for the whole town. The wood from one single tree was enough to build several houses and barns.

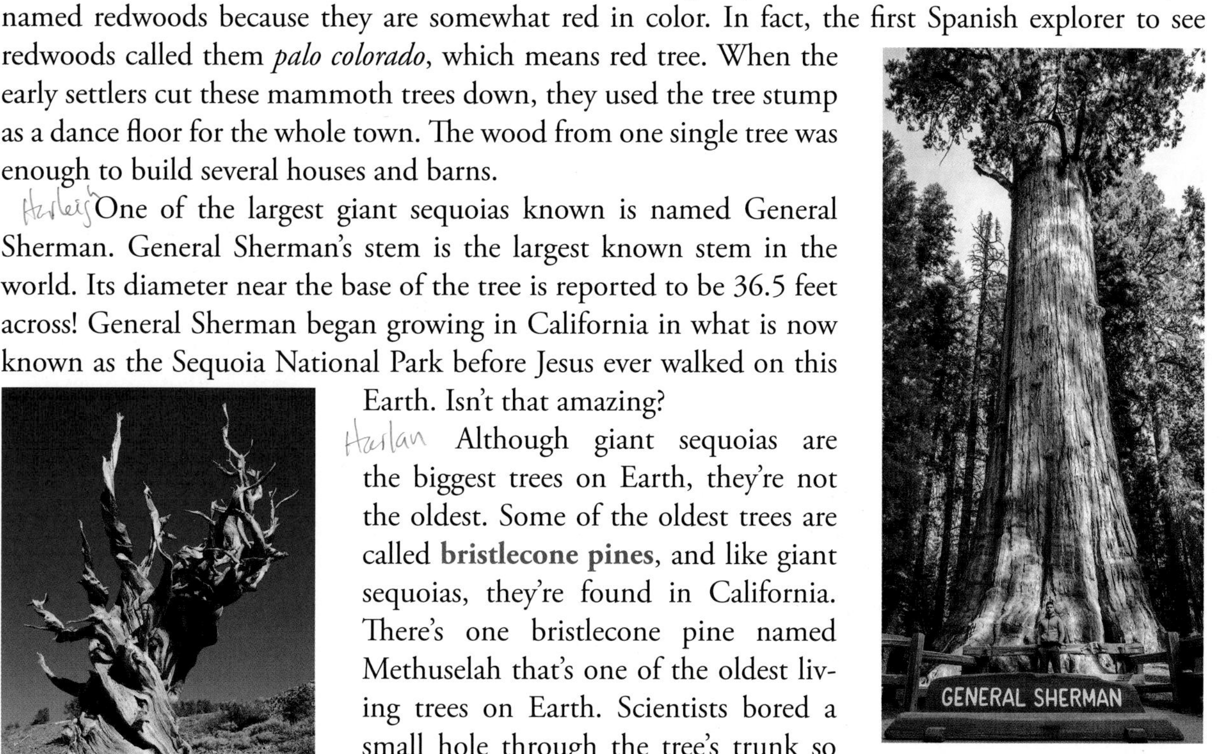

At its base, this giant sequoia measures 36 1/2 feet across.

One of the largest giant sequoias known is named General Sherman. General Sherman's stem is the largest known stem in the world. Its diameter near the base of the tree is reported to be 36.5 feet across! General Sherman began growing in California in what is now known as the Sequoia National Park before Jesus ever walked on this Earth. Isn't that amazing?

Methuselah is a 4,850 year-old bristlecone pine tree.

Although giant sequoias are the biggest trees on Earth, they're not the oldest. Some of the oldest trees are called **bristlecone pines**, and like giant sequoias, they're found in California. There's one bristlecone pine named Methuselah that's one of the oldest living trees on Earth. Scientists bored a small hole through the tree's trunk so they could count its rings. What they found is astonishing—Methuselah is just over just over 4,800 years old! Methuselah was named after Noah's grandfather, who lived longer than anyone else in the Bible.

ACTIVITY 11.1

MEASURE GENERAL SHERMAN

To get an idea of the sheer size of a giant sequoia, let's measure General Sherman's diameter. Using a tape measure and some chalk, go outside and measure 36 feet across. Are you amazed by the size of General Sherman? Perhaps one day you'll go to northern California and see this tree's magnitude with your own eyes! Write down in your Botany Notebooking Journal what you learned from this activity.

SOFTWOOD

Gymnosperm trees are often called **softwood** trees, while angiosperm trees are generally called **hardwood** trees. Why do you think they've been given these names? Which kind of tree do you think would be easier to cut down? Softwood trees are usually much easier to chop down than hardwoods. In fact, thousands are chopped down every year and used as Christmas trees all over the world! Even a child could chop down a small Christmas tree. This is because the wood of gymnosperm trees is generally much softer and easier to cut than that of a hardwood tree. Most furniture is made from pine because of this. Pine is easier to work with when you're cutting and sanding wood to make furniture. It's also easier to lift, carry, and move around. Oak, cherry, and maple furniture are types made from angiosperm trees. This kind of hardwood furniture is heavier and harder to make. Which kind of wooden furniture do you have in your house? Compare how difficult it is to lift one chair compared to another made from a different wood. Try to guess which wood is pine and which is oak, maple, or cherry. Sometimes the manufacturer will stamp the type of wood on the bottom of the furniture. Does your furniture have such a stamp? Look and see!

Gymnosperms are popular Christmas trees because their soft wood is easy to cut.

EVERGREEN

As you may already know, when a tree is green all winter long, we call it an **evergreen**. These trees are different from deciduous trees because they don't lose their leaves in the fall. Which are the most plentiful trees in your area: deciduous or evergreen? Most angiosperm trees are deciduous, while most conifers are evergreen.

Do you remember why deciduous trees must lose their leaves in the autumn? It's because of transpiration. Deciduous trees lose so much water through their big leaves that they would die of thirst in the frozen, dry winter if they kept their leaves. Evergreen leaves are designed to hold water. As a result, they don't transpire very much, so the tree doesn't need to lose its leaves by winter.

However, evergreens actually do lose their leaves, just not all at once. All throughout the year, they lose leaves but new leaves grow back to replace them so they are never completely without leaves. These trees dominate the cold regions of Earth because deciduous trees can't survive as well there. It's hard for them to conserve water high up on a mountain where snow is present many months of the year. Conifers, on the other hand, don't lose much water through transpiration. That's why they're perfectly suited to winter climates and why you'll find them to be most abundant at the top of the globe.

Evergreens like this pine keep their leaves and beautiful color in winter.

However, conifers also grow well in warm, dry regions because of their uncanny ability to hold on to every bit of water they can get.

Explain to someone what you have learned about gymnosperms so far in this lesson.

ACTIVITY 11.2
COMPARE TRANSPIRATION

If your yard has a conifer and a deciduous tree with leaves that are still green, you can do an experiment to measure how much water each transpires in a week. Do you remember when you put a plastic bag over a leaf to see transpiration in action? You're going to do a similar thing in this activity, but this time you'll experiment on outdoor trees.

You will need:
- 2 plastic sandwich bags
- 2 clothespins
- Deciduous tree
- Conifer tree

You will do:
1. Put one sandwich bag around a large leaf on a deciduous tree.
2. Secure the sandwich bag with a clothespin.
3. Put the other sandwich bag around a small clump of leaves on a conifer tree.
4. Every few days, go outside and check the amount of water in each of the bags. You can leave them on the trees all summer long if you'd like.
5. Use a Scientific Speculation Sheet to write down which bag you think will have more water in it. Record whether or not your hypothesis was correct.

CONIFER LEAVES

There are three kinds of leaves found on conifers: needlelike, scalelike, and awl-like. Do you remember I mentioned these in Lesson 6 when we were learning about leaf shapes? Needlelike leaves are smooth and straight and are usually pointy at the tip. Sometimes the needles are long, and sometimes they're short. Scalelike leaves have a bumpy texture that looks like scales. They appear almost like snake skin. These leaves are rough and often have a strong scent that reminds us of Christmas. Awl-like leaves are not scaled, nor are they long, pointy needles. In contrast, they're pointy leaves that look a little like very thin triangles or flat spikes.

Do you have any conifers in your yard or area? If so, go outside and do the activity that follows.

ACTIVITY 11.3

IDENTIFY AND ILLUSTRATE LEAVES

Study the different conifer leaves found on the trees in your yard. In your Botany Notebooking Journal, illustrate each leaf you find. Try to determine whether its needlelike, scalelike, or awl-like. Record what you learn about each type of conifer leaf.

CONIFER SHAPE

In addition to having different leaves than angiosperm trees, conifers are usually shaped differently than angiosperms. Often, they're shaped like a skinny "A." This shape allows a lot of snow to fall off the tree rather than stay trapped in the leaves. That way, the limbs are less likely to break during heavy snows in the winter. The tree's soft wood also allows its limbs to bend down when weighted with snow so that the snow can slide off the tree. If a hardwood tree branch gets weighted down with snow, the branch may become so heavy that it simply snaps off. You can see how God designed conifers in a very special and practical way to survive the coldest parts of the world.

CONIFER CONES

Now remember, the conifer tree is not the only kind of gymnosperm that exists. However, it's the most common, which is why we're concentrating on conifers. Do you remember what separates conifers from other gymnosperms? You're right. Conifers make their seeds in cones. Have you ever seen a pinecone? Most people have. In fact, you've probably held a pinecone in your hand at one time or another. You might even have a pinecone sitting outside in your yard right at this moment! A pinecone is an example of a cone made by a conifer. You'll experiment with pinecones at the end of this lesson, so you'll want to collect one soon.

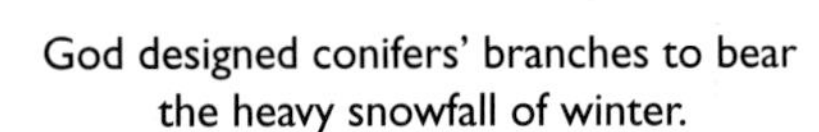

God designed conifers' branches to bear the heavy snowfall of winter.

Each conifer makes two different types of cones: **pollen cones** (which contain pollen) and **seed cones** (which contain eggs). Can you guess which one is the male cone and which is the female cone? From our study of flowers, you probably guessed that the pollen cones are the males, and the seed cones are the females. Both male and female cones usually grow on the same tree. The female cones are by far the most noticeable. They're usually reddish-brown or greenish-yellow when they first begin to grow. After they're pollinated, however, they turn into the dark, woody-looking cone you see on the ground.

Pollen cones are generally softer and harder to see than seed cones. In fact, you might not have noticed them before. Yet without the pollen cones, the seed cones would never develop new seeds. After all, in order for a seed to form, an egg must be pollinated. The pollen cones produce pollen in the

Left: This pollen cone is the male cone containing pollen. Right: Eggs are found in this female seed cone.

spring. When it's ready, the pollen cone bursts open, and millions of pollen grains fly into the air, some landing on the ground, some landing in nearby water, and some reaching their target: a seed cone. After its pollen is released, the pollen cone is no longer needed, so it shrivels up and dies.

The female cone has scales that look something like leaves. These scales will one day be the hard little scales of a pinecone. At the base of each scale are two ovules that contain the eggs. Once the pollen reaches the eggs, they begin fertilization and seeds begin to form. It takes more than a year for the eggs to be completely fertilized. When the year is over and the eggs are fertilized, the seeds begin to mature.

When the seeds are ready, the pinecone is also ready. It begins closed up tight. Then, on a warm and dry day, the pinecone expands, opening its scales so that the seeds can fall out. Each seed is attached to a little wing (often called a key) that helps it float to a new plot of land to grow into a new tree. If the day is cold and wet, the pinecone is designed to stay shut, holding on to its seeds until a warm, dry wind blows in. Isn't it beautiful the way God designed cones to open only when the weather is perfect for a new tree to grow?

BREAKING THE RULES

As you study God's incredible creation, you'll find it often doesn't obey the "rules" human science tries to impose on it. Juniper bushes and trees provide a good example of this. Botanists call junipers conifers. And even though *conifer* means cone-bearer, junipers do not bear cones. Why then are they called conifers? It's because juniper leaves and wood are so similar to conifer leaves and wood. It only makes sense then to group them with the conifers, despite the fact that they don't make cones. To make this even more confusing, botanists insist on calling juniper fruits cones even though they clearly look like berries.

Unlike a normal conifer that produces both the pollen part and the egg part, a juniper will produce either only pollen or only eggs.

Juniper trees are grouped with conifers even though they bear fruits instead of cones.

The yew is considered a conifer even though it produces arils rather than cones.

Plants that behave this way are often labeled male or female based on which part they produce. Can you guess which is which? Of course you know that the male juniper produces pollen and the female produces eggs. Because of this, only the female junipers will bear fruit. Can you explain why? It's because fruit must come from an egg that has been pollinated.

Juniper leaves can be scalelike or awl-like, but one thing is the same with all junipers: they have a strong odor you can smell from a long way off. Junipers are often bushes and shrubs, though there are many kinds of juniper trees as well.

Another conifer rule breaker is the yew. A yew doesn't make cones. It makes an **aril** (ar' uhl), which is a fleshy covering for the seed that looks a lot like a fruit. It's sometimes called a false fruit. Though the arils of a yew are quite poisonous to eat, yews have a natural and beneficial chemical in them called taxol. This chemical has been effective in the treatment of certain types of cancer.

Tell someone what you have learned so far about conifers.

CYCADS

Cycads are gymnosperms that grow in warm, tropical climates.

Even though we've been concentrating on conifers as typical gymnosperms, some gymnosperms are not conifers. These gymnosperms are put in a different category from conifers. One such gymnosperm is the **cycad** (sy' kad). Cycads grow in warm, tropical climates where palm trees are found. They're often mistaken for palm trees because cycads and palms both have a palm-tree trunk topped by a whorl of palmlike leaves, usually without any side branches. The leaves of cycads, like palms, have a central stalk with rows of narrow leaflets on both sides. The similarities between these two plants end, however, when it's time to make seeds.

Palms bear flowers that, when fertilized, make fruits such as dates and coconuts; therefore, palms are angiosperms. Cycads, on the other hand, produce cones in the center of the leaf whorl with male and female cones occurring on separate plants. Because their seeds are made in open containers rather than closed fruits, cycads are gymnosperms. The funny thing about cycads is even though they make cones like conifers, they are not called conifers because their leaves and wood are very different from conifer leaves and wood. The next time you think you see a short, wide palm tree, look to see if there's a cone in the center. You may be looking at a gymnosperm disguised as a palm!

GINKGO BILOBA

The **ginkgo** (geenk' oh) is a most unusual little gymnosperm because its leaves are broad and flat like those of an angiosperm. It also loses its leaves in the fall and regrows them in the spring, which means it's deciduous. Despite the fact that ginkgo leaves are similar to angiosperm leaves, the ginkgo is a gymnosperm because it produces uncovered seeds.

Like junipers, ginkgos produce either pollen or eggs on a single tree. Thus, there are male ginkgo plants and female ginkgo plants. A female ginkgo, like the oak tree, must be at least 20 years old before it can make seeds. When the seed coat comes off its one-inch round, yellowish seed, the fleshy interior creates a terrible odor that smells like vomit. Because of this, people don't plant female ginkgos for ornamental purposes. This makes the male tree much more valuable than the female tree. Before the late 1600s, scientists found fossils of ginkgo tree leaves but never found any living ginkgo trees. The ginkgo was thought to be **extinct**, meaning it lived on Earth sometime in the past but does not exist now. Scientists believed that at some point the plant died out, until a single species, *Ginkgo biloba*, was discovered in 1691 in Japan. Japanese herbalists had been using ginkgo seeds as medicine for hundreds of years.

Though it looks like an angiosperm, the *Ginkgo biloba* is a gymnosperm because it produces uncovered seeds.

Explain what you know about gymnosperms to someone so they can learn what you have.

FORESTS

You've learned a lot about trees, but did you know that most trees are actually found in forests? Forests take up one third of Earth's land. They're found everywhere! Interestingly, there are actually different kinds of forests based on the kinds of trees that grow there.

No matter which kind of forest it is, all forests have three layers. The top layer is called the canopy. It's the uppermost part of the forest where all the tree leaves and branches are exposed to the sun above. The canopy is kind of like the roof of a house; however, some sunlight can pierce through the canopy if it's not too dense. The second layer of a forest is the understory. It's where smaller trees and bushes grow. The bottom layer of a forest is the forest floor. That's where the dead plants and animals decompose and where small plants grow.

Let's explore the different kinds of forests.

TROPICAL RAINFORESTS

We've talked a lot about rainforests. The largest rainforests are in the Amazon, the Congo, and Southeast Asia. These special forests contain the most kinds of plants and animals in the world. When scientists explore these rainforests, they often find new species of animals they didn't know existed before.

Rainforests are usually found near the equator. Even though this is the hottest climate on Earth, the rainforest itself keeps the weather inside the forest rather cool. It's usually between 70–85 degrees all the time. Why do you think that is? Well, imagine all the shade the trees provide. Do you remember how trees release a great deal of water? Indeed, this water fills the atmosphere inside the rainforest making it very humid. Most of the trees found in a rainforest have wide, broad leaves to capture as much of the tiny bit of sun that comes through the thick canopy above as they can. Plants that love shade, such as ferns and

Red-eyed tree frogs are one of the many colorful frogs found in rainforests.

The emerald tree boa makes its home in the tropical rainforests of South America.

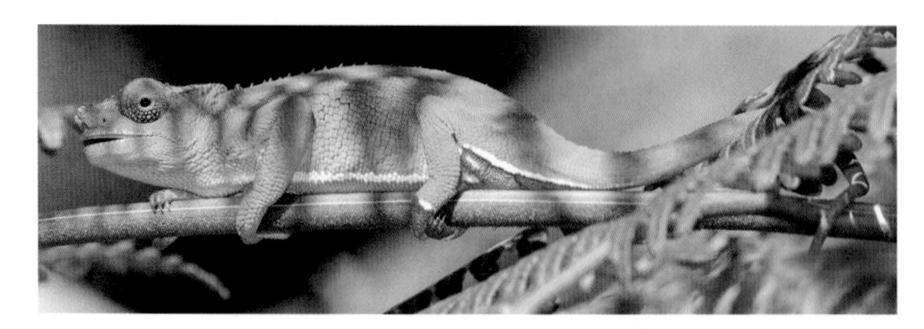

Two-horned chameleons are found in the mountain rainforests of Tanzania.

orchids, thrive in the rainforest. Because so little sunlight reaches the understory or forest floor, many of the animals that live in the rainforest, such as monkeys and colorful birds, are found high in the trees above. Down below, you'll find snakes, colorful frogs, and unusual butterflies.

BOREAL FORESTS

A boreal forest might be considered the exact opposite of a rainforest. These forests are located in the coldest parts of the world, such as northern Canada, Asia, Siberia, Denmark, Norway, Sweden, and Finland. Boreal forests experience long cold winters and short summers. Most of the year, they are covered with snow. Can you guess what kinds of trees are found here? Gymnosperms! In these cold forests, the canopy is extremely thick, allowing little to no light below. As a result, very little grows in the understory or forest floor. The animals found in boreal forests are few and have very thick fur so as to withstand the long, cold winters. Caribou, snowshoe hares, wolves, and grizzly bears are found in these forests.

Boreal forests are home to many grizzly bears.

TEMPERATE DECIDUOUS FORESTS

The word temperate means mild or without extremes, so temperate deciduous forests don't have extremely warm temperatures or extremely cold temperatures year round. As you can probably guess, deciduous forests don't stay green all year long. That's because the leaves change colors and fall off in the cooling autumn weather. These special forests are found in places that experience all four seasons, such as the eastern United States, Canada, Russia, China, and Japan. Deciduous forests contain trees such as the maple, oak, and birch, although they can include evergreen trees as well. Although it rains a lot in the temperate forest, the rain comes throughout each season, including in the form of snow in the winter. Because the leaves fall, providing the earth with plant matter that decays into compost, the deciduous forest's floor is nutrient rich. Many animals live in these forests, including foxes, hundreds of kinds of birds, bears, wildcats, and a great many little mammals.

Temperate deciduous forests provide the perfect environment for the red panda.

TEMPERATE CONIFEROUS FOREST

The spotted wood owl dwells in the trees of the temperate coniferous forest.

Temperate coniferous forests are often found near the coast where the temperature is mild year round, meaning it doesn't get very hot or very cold. These forests are usually made up of conifers that grow very tall. Can you think of a conifer that grows very, very tall? The reason they grow so tall in this kind of forest is because the winters are mild. The plants just keep growing and growing! Although you might find a few deciduous trees here and there, the kinds of trees found in the temperate coniferous forest are mostly redwoods, firs, pines, cedars, and cypresses.

Can you guess where one temperate coniferous forest might be located? You may have guessed the Pacific northwestern of the United States where the giant sequoias live. If so, you are right! They're also found in Canada, southwestern South America, southern Japan, New Zealand, and a few places in northwestern Europe (Ireland, Scotland, Iceland, and Norway). Many animals live in these forests, such as owls, deer, and bears.

FOREST FIRES

Though forest fires are harmful, they can also be beneficial to a forest when controlled.

As you know, forests provide protection, food, and moisture for many species of animals, as well as oxygen for the Earth. However, if a forest wildfire develops, it can destroy thousands of trees in a very short amount of time. In 1988, for example, a series of forest fires swept through a large portion of Yellowstone National Park. Nine of those fires were started by careless human beings. But 42 of them started naturally, by lightning striking dry wood. More than 790,000 acres, about 35 percent of the park, were affected by these fires.

Although forest fires can cause a lot of damage, they can also be beneficial to a forest. In a typical fire, some large trees (especially ones with a lot of moisture in their sapwood) will survive while other trees will burn away. This can actually be a good thing and here's why: the decayed remains of burned trees fertilize the soil. Also, the cleared landscape can become home to a larger variety of plant species than was there before. In fact, based on many observations scientists have made over the years, it has been determined that trying to suppress natural forest fires actually reduces the variety of plant and animal species that are seen in the wilderness. As a result, those who care for our national forests tend to let forest fires burn as long as they pose no threat to people, homes, or buildings. Sometimes during the cool months, forest rangers will conduct a controlled burning of a certain area to promote the good health of that part of the forest. In a controlled burning, certain parts of the forest are set fire and burned down in an organized and controlled manner.

I hope you enjoyed your study of gymnosperms and forests. Next time you see a Christmas tree, remember it is a conifer that was once a tiny seed in a little cone.

WHAT DO YOU REMEMBER?

What are the three kinds of conifer leaves? What is the process by which conifers produce seeds? Why are cycads considered gymnosperms? Describe a cycad. Describe the ginkgo. Why are ginkgos considered gymnosperms? Why do people prefer to plant the male *Ginkgo biloba*? Name the four kinds of forests. How can forest fires be beneficial to a forest?

ACTIVITY 11.4

WRITE A BRISTLECONE PINE STORY

After you've recorded all you remember from this lesson, I want you to create a life story of a bristlecone pine. I want your story to have a parallel story of what was going on in history during the beginning of its life. For example, you might begin the story by saying: "When the pinecone seed rested upon dry ground, it began to grow. At this time, Noah was leaving the ark with his family." You could then write about what happened to the tree and what happened to Noah and his family at the same time. Use the special pages in your Botany Notebooking Journal selected for this activity.

ACTIVITY 11.5

OPENING AND CLOSING PINECONES

Pinecones are programmed by God to open and close according to the outside temperature and humidity. What do you think would happen if you put a pinecone in a bucket of cold water? What do you think would happen if you placed the pinecone in a heated oven? Record your guesses on a Scientific Speculation Sheet.

You will need:

- Pinecone
- Oven preheated to 250 degrees
- Bucket of cold water

You will do:

1. Place the pinecone in the bucket of cold water.
2. Check your pinecone in an hour. Has it changed in any way? Record your observations on the Scientific Speculation Sheet.
3. Place the pinecone in an oven preheated to 250 degrees. Check the pinecone in an hour. How has it changed? Record your observations on the Scientific Speculation Sheet.

LESSON 12

SEEDLESS VASCULAR PLANTS

digging deeper

Those who trust in their riches will fall, but the righteous will thrive like a green leaf.
Proverbs 11:28

What do you trust in most? Some people trust in the things they own to bring them happiness. They think their computer or toys or other belongings are the most important things in their lives. Others believe that true contentment comes from the sport they play or the talent they have. Often, we spend more time worrying about these things than pursuing our relationship with God. But God tells us we must put all our trust in Him. We should look to Him for our fulfillment. Everything else is just icing on the cake. If you set your heart on God, looking to Him for your happiness, you'll thrive like a vibrant green leaf!

SPORANGIUM

You may remember that some plants are vascular and some plants are nonvascular. But do you remember that some plants produce seeds and some do not? There are only a few kinds of vascular plants that don't produce seeds. Instead, they produce sporangia. Do you remember what sporangia are? As a reminder, *angium* means container. So **sporangium** means *spore* container. These seedless vascular plants make spore containers. The spores can grow into new plants;however, they are not seeds.

The little brown dots you see on the underside of these fern leaves are sporangia.

Do you remember what a seed is? A seed is a plant embryo in a protective coat that contains food for the embryo so that it can germinate. In other words, it's a baby plant in a box with its lunch. Do you realize what that means? It means the seed does not need soil or light right away to germinate. Instead, it uses the food stored in the seed to start growing. Spores are not like this at all.

Individual spores are very tiny and, unlike seeds, have no food stored inside them. In order to germinate, spores must have soil (or a suitable substitute) and light right away. They need to begin making food from the very first moment or they cannot germinate or grow. Because of this, a seed is much more likely to grow into a plant than is a spore. Spores need more ideal conditions than seeds.

In this lesson you'll study the vascular plants that produce spores rather than seeds. The most abundant type of seedless vascular plant is the fern.

Ferns are classified in a group of plants called **pteridophytes**. The Greek word *pteris* means wing or feather. You can easily see why ferns are called feather plants. They do look quite like feathers. Let's take a closer look at the anatomy of these feathery plants.

It's easy to see how feathery ferns got their name.

FERN ANATOMY

You can see how a fern is different from the other plants we've explored by looking at the fern's anatomy.

ROOTS

Let's explore ferns from bottom to top. At the bottom, you'll find the roots of the fern. The roots are connected to a rhizome (with ferns, the rhizome is the stem you see). Do you remember what rhizomes are? They're underground storehouses full of nutrients to make a plant grow. Ferns develop rhizomes. Sometimes these rhizomes creep along the ground, forming new rhizomes as they grow. The little fern stems grow up along the ground at each new rhizome, much like grass runners. Other rhizomes are thick and stocky. The fern shoots up above the ground from the rhizomes, forming clusters of leaves. These leaves are what help the plant reproduce. So, a fern's leaf is kind of like the flower. Crazy, huh? Ferns are certainly different than most other vascular plants.

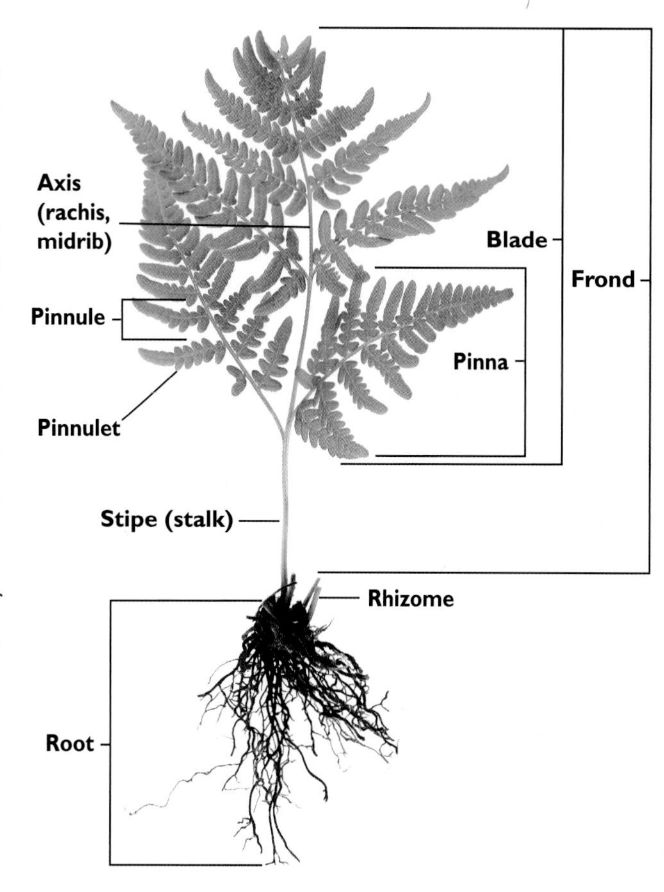

STIPE

The stalk, or **stipe**, is the rhizome of the fern plant. The stipe ends where the leaves begin.

FRONDS

The leafy blade of a fern is called a frond. Do you remember what compound leaves are? You learned about them in Lesson 6. A compound leaf is a leaf made up of many individual leaflets. As you can see, fern fronds are compound and sometimes they can be twice compound.

The center of the frond is called the **rachis** (ray' kus), sometimes referred to as the axis. The rachis is the extension above the stipe, which connects the frond to the central stem of the fern. The leaflets that attach to the rachis are called **pinna** (pin' a). They're often made up of more individual leaflets, called pinnules. Pinnules are like little sub-leaflets off a main leaf. If the frond stem has another stem branching off of it and has its own pinnae, it is twice compound. Study the images below to see the difference between compound and twice compound fronds.

Twice compound fronds are made up of individual leaflets and also subleaflets.

Compound fronds are made up of many individual leaflets.

The little clusters of sporangia you see here are called sori.

SORUS

During certain times of the year, if you look on the underside of a frond, you'll find little bumpy clusters stuck there. Those clusters are the reproductive part of the fern—right there on the leaves!

We call these cluster bumps **sori** (sor' eye), or a **sorus** if we're speaking of only one. The sori are actually clusters of sporangia. As you already figured out, those little sporangia clusters contain the spores that may one day grow into new ferns (if the conditions are just right).

Believe it or not, one single sorus contains thousands of spores. Why is that? Because the conditions must be just right for a new fern plant to grow. Consequently, it's not likely that very many spores will actually become new fern plants. Ferns make up for this by producing a massive number of spores. Although most spores will not grow into new ferns, each fern makes so many spores that at least some of them probably will.

Before we study how ferns reproduce, which is quite interesting, tell someone what you have learned so far about ferns.

FERN LIFE CYCLE

So what does a fern spore need to do to make a new plant? First, the spore must leave its container, the sporangium (sporangium is the singular form of sporangia). When the spores in a sporangium are ready to leave, the sporangium opens up and flings the tiny, dust-sized spores into the air. Because they are so small and light, the spores float in the air until they land somewhere. If a spore lands somewhere that has just the right amount of soil, moisture, and light, it grows into a tiny, heart-shaped plantlet called a **prothallus** (pro thal' us).

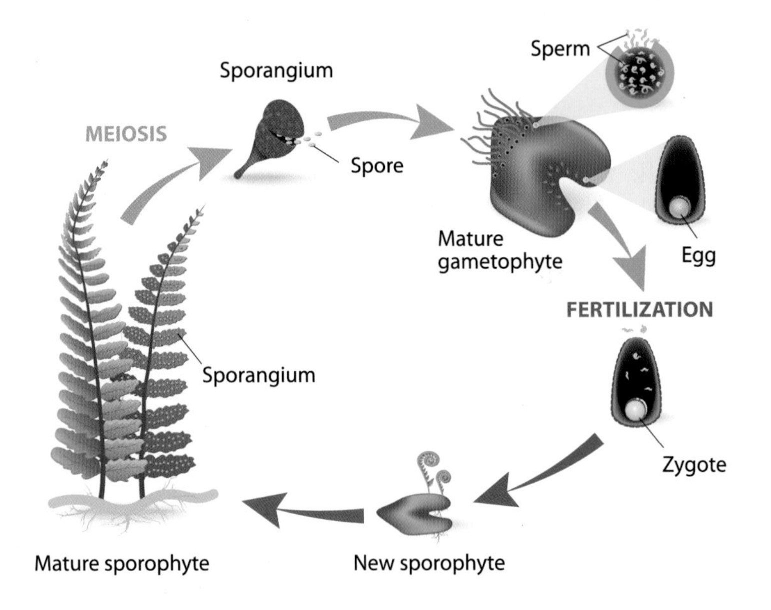

PROTHALLUS

The prothallus is usually very hard to see because it's small and often lightly colored or transparent. If you see a bunch of ferns growing in a forest, it's possible there is a prothallus (perhaps several) growing close by. You would have to search hard with a magnifying glass to find it, however.

The prothallus is not what we think of when we think of ferns. Nevertheless, it's a part of the way that a spore develops into a fern, so we say that the prothallus is a part of the fern's **life cycle**. In a way, the prothallus is like a flower, though it's not nearly as pretty as a flower. Instead, it looks like a little heart-shaped, squishy structure with roots dangling from its base. Those roots are called **rhizoids** (ry' zoyds), and they absorb water and nutrients from the soil.

MALE AND FEMALE

The prothallus is like a flower because it has male and female parts just like a flower has. The male parts on the prothallus are little bumps on its base, and they are called antheridia (an' ther id' ee uh). Do you recognize the first part of that word? Do you remember what the anther on a flower is? It's the male part of the flower. The antheridia are the male parts of the prothallus. They hold the sperm. The female parts of the prothallus are called the **archegonia** (ark' uh guh nee' uh). They contain the eggs.

Once the antheridia are mature, they release sperm to unite with the female part, the archegonia. Believe it or not, these sperm actually swim to find the archegonia. As long as there's a lot of water, the sperm will swim around looking for archegonia to enter. If they find one, they'll go inside and fertilize the egg.

You might think the sperm don't have far to swim since the archegonia are so close to the antheridia on the prothallus, but that's not how it works. Like most organisms in God's kingdom, antheridia must find archegonia from a different prothallus, which means a whole different spore must have landed nearby. You can see how complicated it is for a fern to reproduce.

To make matters even more challenging, the antheridia and the archegonia on each prothallus mature at different times. However, if they did mature at the same time, the sperm might fertilize the same prothallus. They usually need to find another prothallus in order to fertilize an egg. Not only do the spores have to land in close proximity in the perfect weather and soil conditions, but the male part of one prothallus must also mature and send out its sperm at the exact time that the female part of another prothallus is ready to receive the sperm. It's a hard life for a spore.

Once we get past all those timing issues, we have a fertilized egg. Once an egg is fertilized, it can start to grow into what we call a fern. To do this, however, it needs food. Do you know where it gets the food? From the prothallus! After the egg is fertilized, it develops a "foot" that imbeds itself in the prothallus and pulls nourishment from it.

UNFURLING OF A NEW FERN

As the new fern finally begins to develop, it first grows into what is often called a **fiddlehead** (shown in the picture to the left). Can you see why it's called a fiddlehead? It looks like the top of a violin, which is often called a fiddle. When the fiddlehead is ready, it unfurls and develops into the beautiful frond that we recognize as a fern.

Do you see why I wanted you to understand that a spore is definitely not a seed? Think about all the steps a spore must go through to become a new fern. First, it must develop into a prothallus. Then, it must release sperm. If the sperm of one prothallus swims over to meet an egg from another prothallus, fertilization will occur. The fertilized egg then starts to develop into what we think of as a fern, but only when it starts pulling nutrients from the prothallus

This fiddlehead is unfurling, revealing the leaflets of the frond.

that held the egg. This entire process is actually a lot more detailed than I've explained here. It's called an alternation of generations life cycle, and you'll learn a lot more about it when you take biology in high school.

Can you explain to someone all that you learned about fern reproduction?

ALTERNATIVE REPRODUCTION

After learning how complicated it can be for a spore to reproduce, you might wonder why there are so many ferns on the forest floor. Interestingly, God created alternative ways for ferns to make new plants.

As I mentioned earlier, ferns can grow new fern plants by spreading runners along the ground from their rhizomes. The runners grow new fern plants at each new rhizome location. However, it's not really a new plant, is it? No, each new fern is a clone of the original plant. It has the exact same DNA as the plant that made the runner.

Another method of reproduction is self-rooting. When fronds grow thick and heavy, the frond of some ferns leans down and touches its tip into the dirt. Guess what? The fern can sprout new ferns right there from the tip touching the dirt! This is a bit like rooting a plant in soil. But again, those new plants are clones of the original Some fronds even grow miniature fernlets that fall to the ground and start a new plant.

There is yet another way that some ferns can reproduce. When they're in the prothallus stage, the prothallus can make a little packet of cells called a **gemma** (jem' uh). The gemma can then be carried off by animals or water to form a new prothallus somewhere else. With all the different ways ferns can reproduce, is it any wonder you find them on nearly every forest floor?

Explain to someone the alternative ways a fern can reproduce.

PTERIDOMANIA

Have you ever been obsessed with a toy, book, or idea? I once knew a boy who was obsessed with the Titanic. I knew a girl who could never get enough of birds. She knew everything about birds and her whole room was decorated in a bird motif.

Because of their unique beauty, ferns have been a popular plant among gardeners for thousands of years. But at one time, they became even more than a special plant to grow; they became a national obsession! In England, from the 1850s to the 1890s, ferns were so wildly popular that most everyone was driven by fern madness. Young and old, rich and poor, everyone became fanatical about ferns. They called this 50-year period of fern obsession pteridomania (mania means madness).

This oil painting titled *The Fern Gatherer* depicts the pteridomania era when fern gathering was wildly popular.

People participated in fern hunting parties during the time of pteridomania.

Dad

In the beginning, ferns were considered plants that only very intelligent people liked or understood. Then, out of nowhere, ferns became everyone's hobby. Lots of books about ferns were published. People collected all varieties of ferns. The more ferns you had, the more special you felt. Fern hunting parties were arranged in the wild to find new species. People built special glass aquariums for their ferns. Some people would build a large glass nursery called a fernery in their yard. Others simply grew ferns all over their property. When visitors came, people would proudly display their many, many different ferns. The rarer or more unique the fern, the better. Fern stands were important pieces of furniture in homes during that time, and the fern motif became the most popular design on everything. Paintings, furniture carvings, upholstery, wallpaper, rugs, tea sets—you name it, the item was decorated with ferns. The craze spread from England to all of Europe and even America!

Let's take a page from the pteridomania time in history and create some fern artwork for our wall.

ACTIVITY 12.1

CREATE FERN ARTWORK

You can create beautiful artwork by pressing a fern covered with paint onto a blank sheet of paper. These fern imprints can be framed and used as very stylish home décor! You can also choose fabric paint and make fern hand towels. If you prefer, use the method from the leaf activity in Lesson 6 and decoupage real ferns to hang on the wall.

You will need:

- Paper
- Several fern fronds
- Several colors of paint
- Paintbrush or sponge

You will do:

1. Use the paintbrush or sponge to paint the entire fern frond on one side.
2. Turn it over onto a piece of paper while the paint is still wet.
3. Place another piece of paper on top of the frond and carefully pat the entire frond.
4. Remove the paper and the fern frond with great care. You now have a beautiful fern imprint!
5. Do this with other fern fronds and colors to create beautiful fern artwork.

TYPES OF FERNS

Ferns can be a few inches tall or up to 12 feet tall. Some ferns need a warm tropical environment to grow. Others stay green and lush even through the hardest winter. Some fern varieties can live up to a hundred years. That's an old fern. Although there are thousands of different kinds of ferns, we're going to take a look at just a few to see the variety of ferns God made.

BOSTON FERN

Do you have a fern plant in your house? It's probably a Boston fern, *Nephrolepis exaltata*. These ferns have bright green leaves that arch outward as they grow. They're perfect indoor plants because they like temperatures between 65 and 75 degrees; however, they do prefer a sunny spot in the house. Studies have found that Boston ferns do an amazing job of cleaning the air in your home. They remove all kinds of toxins. So if you're looking for an ideal plant, the Boston fern might be the winner.

Boston ferns are popular and beneficial houseplants.

LADY FERN

The most common fern in North America is the lady fern, *Athyrium filix-femina*. This fern grows naturally and wild in the damp areas of shady woods. It's also commonly grown as a houseplant or a garden fern. The lady fern is lighter green than the Boston fern. The fiddleheads on this are quite unusual because they are red before they unfurl!

Lady ferns like this one make lovely garden ferns.

STAGHORN FERN

These ferns have the oddest leaves. Some look like the horns of a deer. The sporangia spread over the edges of the fronds, almost completely covering the frond. There are 18 different species of staghorn ferns. They are part of a group of ferns called *Platycerium*. All of these ferns have uniquely shaped leaves. They are considered tropical plants because they grow naturally in tropical areas such as the Philippines, Southeast Asia, Indonesia, Australia, Madagascar, Africa, and America. Although staghorn ferns grow wild in the tropics, gardeners love to grow them as well. If you live north of zone 10, these ferns must be brought indoors during the winter if you want to survive.

The uniquely shaped staghorn fern grows natural in tropical areas.

CHRISTMAS FERN

The Christmas fern, *Olystichum acrostichoides*, is a favorite in gardens because it's an evergreen up through zone 3. It also grows very well in all kinds of conditions, from dry to moist. During the winter, the fronds lay flat on the ground. However they perk up when spring comes. I've bordered my entire garden with Christmas ferns to ensure I have something green to look at all winter long. As long as there's shade, your garden will reproduce lots of Christmas ferns year round.

Christmas ferns are beautiful additions to a home garden.

Most of these clover-shaped ferns prefer to grow in watery places.

Harleigh

CLOVER FERN

The clover fern is the most unusual fern because its leaves are shaped like a four-leaf clover. That means it really don't look like a fern at all. Most clover ferns, called *Marsilea*, prefer to grow in very watery places. Because of this, they're often called water clovers. However, there are some species that can grow in the driest of places. Many people use this fern as a ground cover, like grass.

TREE FERN

Harlan

Tree ferns are giant ferns that look like trees. In fact, tree ferns even have trunks, but they are not made of wood like normal tree trunks. Instead, tree fern trunks are composed of a mass of roots that intertwine, giving support to the fronds. As you can see from the picture below, they look a lot like palm trees and cycads (the gymnosperms you learned about in the previous lesson). When you see tree ferns, you may not be able to tell them from palm trees or cycads. However, if a tree fern happens to be in the right stage of its growth, you might notice some fiddleheads near the top. That would be one way to tell tree ferns apart from palm trees or cycads. There's a better way to determine whether you are looking at a palm tree, cycad, or tree fern. You can investigate how the plant reproduces. Palm trees reproduce by making coconuts; cycads reproduce by making cones; and tree ferns reproduce with spores.

Interestingly, tree ferns have trunks made of masses of roots to hold them up.

Dad

There are more than 10,000 species of ferns. So next time you are out in the woods, keep a lookout for ferns!

Well that about does it for vascular plants. You've already learned so much this year. What was the most interesting thing you discovered about ferns?

WHAT DO YOU REMEMBER?

What kind of roots do ferns have? What is a frond? What are the spots sometimes found on the underside of the frond? In the fern life cycle, what is the little structure with male and female plant parts? Can you name the male part? Can you name the female part? What must be present for the sperm to reach the egg? What is the little baby frond that unfurls called? What are some other ways ferns produce new plants? What are tree ferns? How are they different from cycads and palms?

ACTIVITY 12.2

ILLUSTRATE A FERN

After recording in your Botany Notebooking Journal what you've learned in this lesson, make an illustration of a fern frond. Next, make an illustration that shows the life cycle of a fern. You can use the illustration on page 180 as a guide.

ACTIVITY 12.3

BUILD A SMALL FERN TERRARIUM

Let's create a fern display like the people of the pteridomania era!

You will need:

- Large glass container with a lid
- Pea-Ver-Comp mixture
- Miniature fern

You will do:

1. Fill the bottom of your container with the soil mixture.
2. Plant your fern in the soil.
3. Water your fern well.
4. Put the lid on your container.
5. Voila! You just created a fern terrarium!

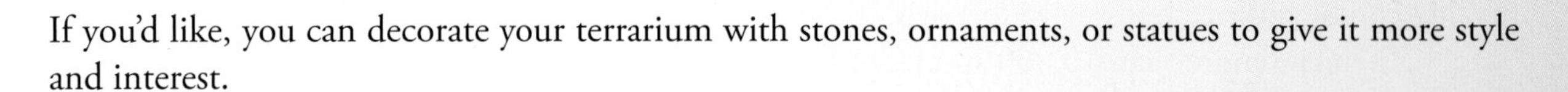

If you'd like, you can decorate your terrarium with stones, ornaments, or statues to give it more style and interest.

LESSON 13

NONVASCULAR PLANTS

digging deeper

The Lord is my shepherd; I have what I need.
He lets me lie down in green pastures;
he leads me beside quiet waters.
He renews my life,
Proverbs 11:28

Do you ever find yourself worrying? In our lives, there are so many things we could worry about. But God wants us to be at peace. He wants to be our Shepherd and lead us to places of rest. Imagine lying down in a peaceful green meadow beside a quiet stream. That's how God wants you to feel every day. Jesus came to bring you that kind of peace. I hope you'll put your worries aside and let Him care for you. Only then will you experience the kind of soul rest the Shepherd promises in Psalm 23.

BRYOPHYTES

Do you remember the difference between nonvascular and vascular plants? Explain the difference in your own words. While vascular plants have liquid-carrying tubes in their bodies, nonvascular plants do not. Unless you live in an area that has little moisture in the air, you probably have nonvascular plants in your yard. They're likely growing on trees, rocks, and maybe even on your house! What plants am I talking about? I'm talking about mosses. Botanists put mosses in a scientific category called **bryophyta** (bry oh fie' tuh). Do you remember what the Greek term *bryo* means? It means moss. It makes sense, then, that mosses are called *bryophyta*, which means moss plants.

There are many different kinds of nonvascular plants that are not mosses. However, as a group, nonvascular plants are called **bryophytes**. So let's dive into the study of these interesting plants.

Mosses are nonvascular plants called bryophytes.

Bryophytes have leaflike structures, but they are not leaves. They also have stemlike structures, but they are not stems. In addition, bryophytes have rootlike structures underneath called rhizoids, but they aren't true roots. Why then do these structures look like stems, leaves, and roots when they are actually not? Because in a bryophyte, these structures don't have xylem and phloem, those inner tubes that carry fluid.

Because bryophytes don't have stems, roots, and all the other wonderful fluid-carrying features of vascular plants, they cannot get their water from the ground. After all, roots must send the water they absorb from the ground to the other parts of the plant. Without tubes to carry the water, nonvascular plants can't do that. As you learned in Lesson 1, nonvascular plants act like paper towels, absorbing water and nutrients that will soak through to other parts of the plant. This process is called diffusion. Because nonvascular plants must use diffusion to transport fluid, nonvascular plants can

Though they look different from mosses, liverworts are also in the nonvascular bryophyte group.

never grow very tall. If they grew tall, water would not be able to soak through to all parts of the plant. That's partly because gravity would keep the water pooled at the lowest part of the plant. So bryophytes are, in general, the smallest members of the plant kingdom.

Nonvascular plants are like vascular plants in one very important way: they make their own food using chlorophyll, sunlight, water, and carbon dioxide. Do you remember what that process is called? That's right! It's called photosynthesis. Of course, the fact that moss is green should tip you off to this fact. It's the chlorophyll used in photosynthesis that gives moss its green color.

MOSSES

Is there moss growing on trees near your home? Sometimes you'll find little patches of moss on the ground, on wooden structures like homes and fences, and even on rocks. Mosses flourish in wet environments. Remember, they don't have roots that can grow in search of water, so they must absorb water directly through their bodies. A wet environment, therefore, is ideal for them. Usually, moss can be found growing more abundantly on the north side of trees (in the Northern Hemisphere). Do you know why this is? Because the sun doesn't hit this side of the tree directly. That means it doesn't dry out as much as the rest of the tree.

Despite this, you *can* find mosses in very dry climates, even deserts. The reason is mosses have the interesting ability to **desiccate** (des' uh kayt). This means they can dry out for long periods of time without actually dying. Without water, mosses will turn brown or yellow because they cannot perform photosynthesis. But they will not be dead. Even after a long time without water, as soon as it rains, mosses will immediately start photosynthesis and turn a lush green again.

You can see that moss is flourishin in this wet environment, covering every rock and tree in sight.

USES FOR MOSSES

Mosses are important to creation in many ways. They provide a home for lots of tiny creatures and are useful to birds for building nests. Most animals don't eat moss because it doesn't have a lot of nutritional value. However, some animals such as bears, deer, and turtles will eat it if they can't find any other kind of food. There is one animal, however, that eats it regularly: the reindeer. Reindeer live in cold climates, and moss has a special chemical that keeps the fluids inside the reindeer from freezing, even on the coldest of days. Moss, then, is a kind of reindeer antifreeze!

God provided for the health of reindeer when He created moss!

Of course, there's a type of moss you should be very familiar with: peat moss. Peat moss can absorb a lot of water. Because of this, it's used quite a bit as packing material and mulch. It's called peat moss because as it grows in the wild, it decomposes into a peat, which is an excellent kind of soil in which to grow plants. Peat also burns well when you dry it out, so it can be used as fuel. In fact, several electric factories in Ireland burn peat to generate electricity.

Peat moss is a very valuable bryophyte because of its many uses.

Tell someone all that you have learned about moss before moving on to learn about reproduction and other bryophytes.

MOSS REPRODUCTION

Just like ferns, mosses produce sporangia. What does that mean? It means they make spores. Since nonvascular plants produce spores like ferns do, you might expect their life cycles to be complicated, like the fern lifec ycle. Well, you are right. Since mosses are the most common nonvascular plants in creation, I want to use moss reproduction to illustrate how all nonvascular plants make new nonvascular plants.

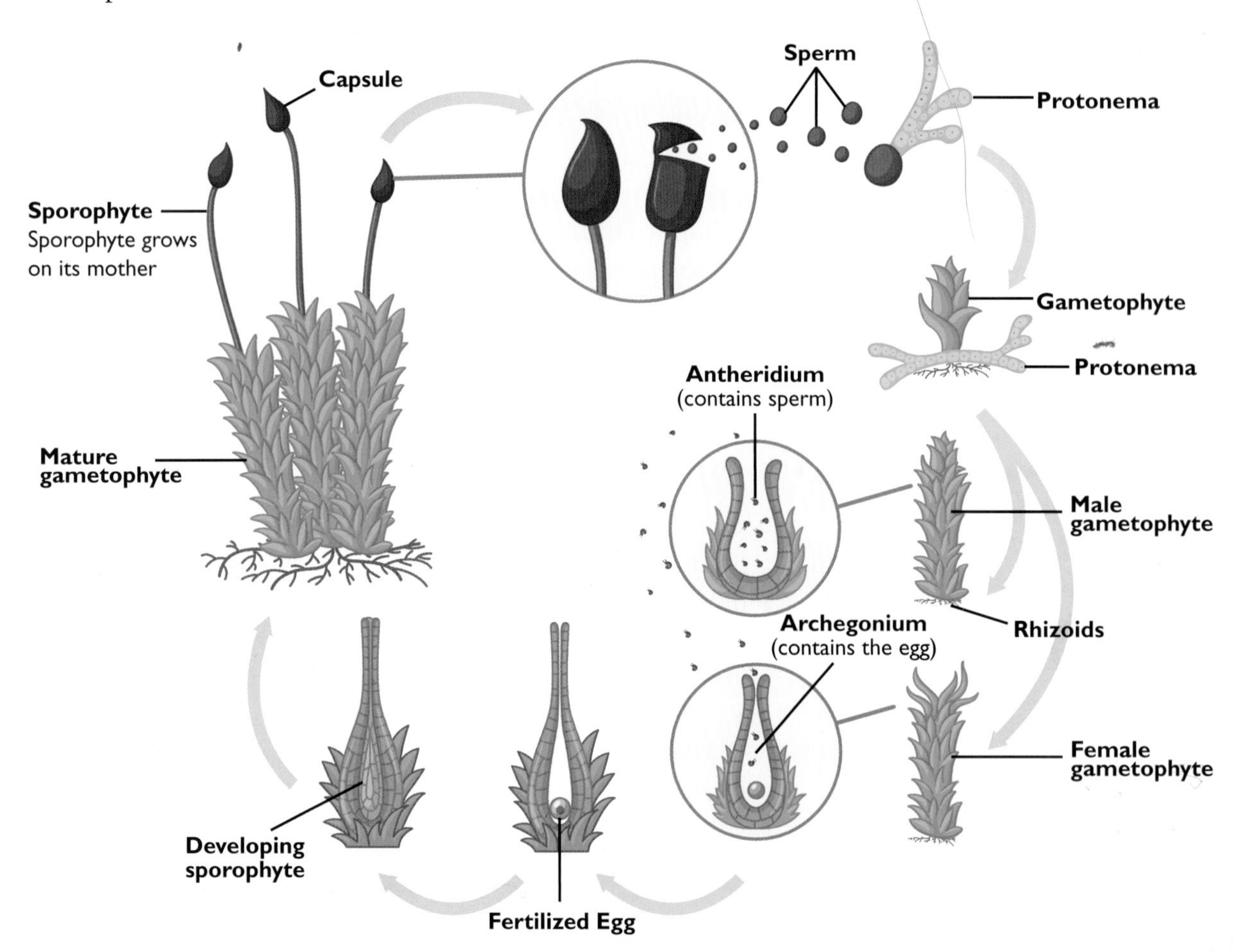

Tiny moss plants grow from spores that germinate in the ground. The really complex part, however, is how these plants form new spores. You see, there are male mosses and female mosses. Guess what male mosses have. They have little structures called antheridia. Do you remember that term from the previous lesson? Can you tell me what a moss's antheridia produce? Yes! They produce sperm. How do you think the sperm get around? That's right. They swim, just like fern sperm. You can see how important it is that the environment is moist. Otherwise, new moss plants wouldn't have a chance.

Can you tell me what structures female moss plants have? They have archegonia, which contain eggs. As you have already guessed, the sperm from moss antheridia swim to the eggs in moss archegonia so that fertilization can take place. Once fertilization happens, a little stalk is produced right inside the archegonium (that's singular for archegonia). We call that stalk a **sporophyte** (spoor' uh fyt). Do you see the sporophyte in the diagram on the facing page It grows upward, and guess what it produces? Spores! Because it's a little stalk pushed upward from the plant, it can disperse the spores farther away from the original plant.

Like ferns, mosses have other ways they can reproduce. Bits of a single moss plant, if broken off, can grow into a completely new moss plant. Of course, they are clones of the original. You know all about that, don't you? Some mosses also produce little structures called **brood bodies**. These structures are designed to separate from the plant then grow into a different moss plant. These are also clones.

Do you see the tiny structures sticking up out of the moss? They're called brood bodies and God designed them to grow into new moss plants.

Describe moss reproduction in your own words.

LIVERWORTS

Have you ever heard of liverworts? You probably have but didn't notice. They are quite small, unnoticeable plants that look a bit like flattened moss.

Liverworts were named in England many years ago. You see, many species of this nonvascular plant have a shape that looks like a person's liver. Thus, people used what was at the time their word for **liver (lifer)** and their word for **plant (wyrt)** to come up with the name liverwort. Back in those days, people believed that if a plant was shaped like a certain body part, it must help that part of the body. So guess what ancient people used liverworts for? They used them to try to treat diseases of the liver. Of course, today we know that such an idea is not correct. As far as medical science can tell, liverworts have no serious medicinal value.

If you want to search for liverworts, look in shaded, moist areas, such as on the ground beneath shrubs or on a shady stream bank. There are two main

These leafy liverworts form structures that look like leaves.

types of liverworts: **leafy liverworts** and **thallose liverworts**. Leafy liverworts look like leaves on a stem lying horizontally on the surface of the ground. The "leaves" are most often arranged in two rows, but in many species there is a third row of much smaller "leaves" that can be seen with a magnifying glass. Now remember, since these plants do not have vascular tissue, they really don't have leaves and stems. They just form structures that look like leaves and stems. The majority of liverworts are leafy. Thallose liverworts do not appear to have leaves or stems. They tend to look like a mass of flattened or wavy green tissue on the ground. One thing that's quite interesting about thallose liverworts is that the sperm are often found inside a special cup-shaped area where a raindrop would splash. The sperm then uses the raindrop like a little car, transporting it toward the archegonia.

Thallose liverworts can be identified by their little cuplike structures containing sperm.

Explain aloud what you have learned so far about liverworts.

ACTIVITY 13.1
HUNT FOR MOSS AND LIVERWORT

It's time to gather your nature journal and start hunting down some moss and liverwort to illustrate. You'll probably need a magnifying glass. Go outside and begin your search by looking on the north side of the trees in your yard. If you live in a dry climate, take a trip to a swamp, creek, stream, lake, or other watery area. Once you've discovered some of these little plants, try to draw them in your nature journal. Don't forget to put the time and date!

LICHENS

Let's conclude this lesson with a very special nonvascular structure that isn't at all like moss. These little leafy structures are called lichens. Lichens are not classified as plants. However, they are often discussed with the nonvascular plants because scientists used to think they were nonvascular plants. After studying lichens for some time, scientists have come to the conclusion that lichens are really made up of two different kinds of living things: fungi and algae (al' jee). Do you know what a fungus is? A **fungus** is a special kind of living organism that feeds off dead creatures. We call them decomposers because they decompose (break down) dead matter so that the organic matter can be recycled back into creation. A mushroom is an example of a fungus. We'll study all about fungi in the next lesson. So, what is an alga? An **alga** is a microscopic organism that typically lives in water and performs photosynthesis. They

Lichens are a mix between fungi and algae.

have chlorophyll, so Algae make their own food. But they aren't exactly plants. Some also consume other organisms the way animals do. So algae seem to be a mix between a plant and an animal, but they look and behave more like a plant in most cases. Lichen is mix between these two groups of organisms. Does that make lichen a plantish animal? I'm not sure. But it's something of the sort.

Here's the way this plantish animal forms: when a certain kind of fungus and a certain kind of alga meet, if they're a good match and conditions are right, they will form a union and create lichen! The alga makes the food through photosynthesis and the fungus absorbs the water and nutrients they both need. They are a good pair, aren't they? When two different organisms form a relationship in which they are dependent on one another, it's called **symbiosis** (sim by oh' sis). Lichen is an example of symbiosis between algae and fungi. Can you think of another example of symbiosis that we have discussed in this book? Remember the process of pollination? Flowering plants (angiosperms) and their pollinators live in symbiosis. The plants provide food (nectar) for the pollinators, and the pollinators help the plants reproduce.

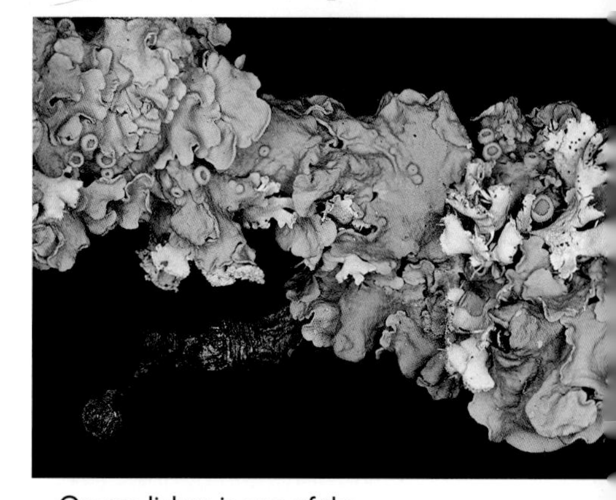
Orange lichen is one of the many unusual varieties of lichen.

Lichens come in all shapes, sizes, and colors. They can look like rust-colored dye that has been painted onto a rock or tree; a scalloped, wrinkled gray or green sheet; lacey pads; bushy clumps; chaotic hair strands of black, gray, or green; and even branching, tiny, gray structures that look like deer antlers! The most common lichen is the kind found on trees. It's usually green or gray and looks like wrinkled leaves growing on the trunk of the tree. If you look hard, you can probably find lichens growing near your home.

Lichen often grows on tree trunks, benefitting the tree by providing an additional layer of protection.

In some environments, lichens can literally cover a tree trunk and all its branches. Believe it or not, this does not harm the tree at all. You see, even though the lichens are growing on the tree, they are not taking anything from the tree. They don't need to take food from the tree because the algae part of the lichen makes the food. They don't even take water from the tree. They absorb water from the air, and they absorb the rain that falls on them. If anything, the lichens help the tree by giving it another layer of protection besides its bark.

When lichens are wet from rain or dew, they grow actively. If they dry out, they stop growing. Like mosses, however, lichens do not die when they dry up. They just lie dormant until the next rain when they can begin growing again. Do you remember what it's called when a plant dries up but doesn't die? Desiccate. Lichens, like mosses, desiccate.

These interesting British soldier lichens are named for their bright red fruiting bodies.

POLLUTION MONITORS

Remember, lichens absorb nutrients and water directly through their tissues. They don't use roots that are imbedded into the ground like other plants do. As a result, they tend to absorb not only the water, but nutrients and other substances that are in the air and water, including any pollution that is there. When the air around lichens is clean, the lichens thrive, but when the air is filled with too much pollution, the lichens cannot grow and will become grayish or even black. Often lichens will simply die off as the air becomes more and more polluted. A city with very poor air quality will have few lichens on the trees.

This particular lichen is very sensitive to pollution and will die if the air quality becomes poor.

Scientists and engineers spend a lot of time designing and building devices to measure how clean the air is in a particular area. More than a hundred years ago, however, a Finnish scientist named William Nylander noticed that lichens on trees in the country around Paris were not found on the trees within the city. He assumed that if they grew in the country around Paris, they should grow in the city as well. Because this was not the case, he concluded the lichens must have been killed by the air pollution being produced by all the new factories in Paris.

These days, we understand that lichens are, indeed, a great measure of air quality. Since they tend to soak up everything that is in the water and air that touches them, even pollutants, they are very sensitive to all of them. When there are too many harmful chemicals in the air, lichens die. If you live where there are many lichens, it probably means the air around you is clean. If there are only a few lichens in your neighborhood, the air you are breathing is probably not as clean as lichens prefer it to be. But you could also be living in a very dry climate. Remember, lichens need lots of moisture.

USES FOR LICHEN

In addition to helping protect trees and giving people a good monitor of the air quality, lichens are important to our environment, just as other plants are. They are a source of food for both caribou and reindeer. They are also homes for spiders, mites, lice, and other insects. Hummingbirds often use lichen (along with other things like moss and spiderwebs) to build their nests.

Lichen is especially useful to hummingbirds when building their nests.

People have learned to use lichens as well. Because lichens come in many different colors, some use them to make dyes for fabric. People also eat particular varieties of lichens, although other varieties are poisonous. One species of lichen, called wolf lichen, was actually used by Native Americans to poison wolves. They also used it to make poison in which they could dip their arrowheads. Some lichens were even used as medicine. Recent studies have shown that lichens do, indeed, contain antibiotic chemicals that help in the treatment of certain kinds of infections.

This poisonous lichen got its name from Native Americans who used it to poison wolves.

Lichens grow very slowly, usually less than one inch in an entire year. If you find lichen growing on a branch that has fallen, you can collect some for your nature journal. However, it's best not to peel it off of a tree.

You've now covered nonvascular plants and the special organism called lichen. You are truly becoming an expert botanist, and I'm proud of you for getting this far in your studies. Good job!

WHAT DO YOU REMEMBER?

How do nonvascular plants distribute moisture and nutrients? What does it mean when a plant desiccates? Where do the sporophytes grow once a moss plant is fertilized? Describe the two types of liverworts. What two organisms combine to make lichen? What are some uses for lichen?

ACTIVITY 13.2

ILLUSTRATE THE MOSS LIFECYCLE

In your Botany Notebooking Journal, illustrate the moss lifecycle using the drawing on p. 190 as a guide. In addition, explain the entire lifecycle. Then write down all you remember about bryophytes.

ACTIVITY 13.3
BUILD A LICHENOMETER

If you live in a climate that has regular moisture in the air, you'll be able to discover how clean the air is in your area by the rate of lichen growth on the trees. To do this, you'll need a lichenometer. Let's make one now!

You will need:

- Coat hanger
- Yarn
- Tape
- Trees with lichens on them, preferably in different places (Lichens prefer oak trees because they can keep the water locked into the deep ridges of the oak bark better than in the bark of most trees.)

You will do:

1. Bend the hook of the coat hanger until it forms an oval. Cover it with tape for safety.
2. Bend the entire coat hanger into a long oval or rectangle shape.
3. Cut several pieces of yarn about 10 inches long.
4. Cut three pieces of yarn about 20 inches long.
5. Begin tying the yarn across the bent coat hanger to make a series of squares out of the yarn (see the picture on the right).
6. The end product should resemble a tennis racket. (Of course, if you have an old tennis racket, you could just use that instead.)

Now that you've made your lichenometer, you can use it to measure the amount of lichen on trees in various locations. Look for a particular type of tree (oak trees are best) that appears in lots of different places in your area of the country. Find the place on the tree with the most lichen growth and place the lichenometer over that area. Count the number of squares that are completely filled with lichen. If a square is only partly filled, don't count it. Take the number of squares that are completely filled with lichen and divide that number by the total number of squares on your lichenometer. Then, multiply by 100. The result will be the percentage of your lichenometer that was filled with lichen. For example, if my lichenometer had 40 squares in it and 16 of them were completely filled with lichen, the percentage would be:

$$16 \div 40 \times 100 = 40\%$$

With this lichenometer, you can determine how clean your air is! Of course, this isn't a perfect formula. Obviously, lichen grows more abundantly in humid areas than in dry areas. But it's a great start to discovering the air quality in your city.

ACTIVITY 13.4

CREATE MOSS GRAFFITI

Have you ever considered decorating with graffiti? You can actually make natural graffiti using moss! You'll create moss paint and your own design, whether your name, a picture or anything you can think of. This project takes 6 weeks to complete. In the end, I think you'll be excited with the results!

You will need:

- 3 handfuls of moss
- 2 tablespoons of water retention gel
- Half cup buttermilk
- Blender
- Bucket
- Paintbrush (2 inch)
- Spray bottle with water
- Piece of plywood or another surface where you want your moss to grow (Ideally you want the surface to be located in a shady area.)

You will do:

1. Rinse the moss with clean water.
2. Pull it apart and put it in the blender.
3. Add the buttermilk and water retention gel.
4. Blend until the mixture is smooth and creamy.
5. Pour the mixture into the bucket.
6. Use the paintbrush to apply the moss paint to your chosen surface in the design you wish it to grow.
7. Spray your design every other day with water and watch it grow.
8. Be sure to photograph it for your nature journal!

LESSON 14

MYCOLOGY

digging deeper

For I know the plans I have for you," declares the Lord, "plans to prosper you and not to harm you, plans to give you hope and a future." Jeremiah 29:11

God has special plans for everything and everyone. In this lesson you'll learn about some of the special ways He created fungi to help plants and people. These unique organisms have a purpose. But did you know that God also created you for a purpose? Yes, indeed. You are much more important than the plants and animals because you were made in the image of God Himself. You have His Spirit within you. And guess what? He has great plans for your life! You can always trust that God is for you, that He's on your side, and that He loves you completely and fully, no matter what you do. As you come to the end of your science studies for the year, embrace God's plan for your life. Believe that He desires good things for you and for your future.

FUNGI

Well, here we are! We've come to the final lesson of our science course. You are done studying botany and are now about to embark on the study of **fungi** (fun-geye). I think you'll find this chapter quite interesting!

You've probably seen fungi in your outdoor explorations this year. The study of fungi is called **mycology**. The Greek word *myco* refers to mushrooms. And of course, all the little mushrooms you see popping up from the ground are fungi. We're going to focus our fungus study on mushrooms, but we'll take a peek at the other kinds of fungi as well because they are quite fascinating and important.

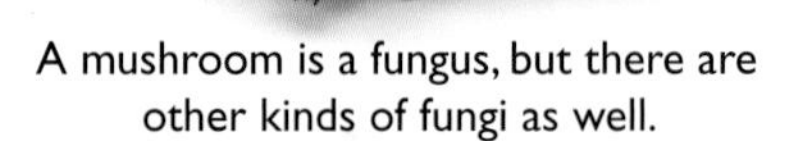

A mushroom is a fungus, but there are other kinds of fungi as well.

So where do we get the word fungi? It comes from the Latin word for mushroom, but it actually originates from the Greek word sphongos. That sounds a bit like fungus, which is the singular form of fungi. However, the reason the Greeks called them this is because *sphongos* means sponge in Greek. The Greeks apparently thought mushrooms looked a bit like sponges. Look at the image on the left. Can you see why the Greeks might have thought this?

We all know mushrooms are not sponges. So what are they?

CONSUMERS

This mushroom is not a plant because it cannot make its own food.

Because they grow out of the ground like plants, scientists once thought they and other fungi were a kind of plant. For many years, studying fungi was included in the study of botany. But now scientists know that fungi are not plants.

Can you figure out why fungi are no longer considered plants? What unique thing does a plant do that no other living thing does? I hope you said, "makes its own food." If you did, you got it right! While plants are producers, fungi are consumers. What do fungi consume? Lots of things. But mostly they live off plants and animals, dead or living, depending on which kind of fungus it is.

Let's learn about the different kinds of fungi.

FUNGUS AMONG US

There are three types of fungi: yeasts, molds, and mushrooms. In each group there are types of fungi that are very beneficial for plants and animals. But each category also has types of fungi that are quite harmful.

Let's take a brief look at yeasts and molds before we jump into the mushroom patch.

YEASTS

Yeasts are microscopic, egg-shaped, unicellular fungi. Unicellular means that the whole organism is contained in only one cell. The prefix *uni* means one. Have you ever heard of a unicycle? It's a bicycle with one wheel. Interestingly, yeasts are kind of like that. They only have one cell and are so small they can only be seen with a microscope.

There are around 1500 species of yeasts, but most of them like the same kind of food. If you could guess what kind of food most yeast enjoy, what would you guess? Well, like many organisms on Earth, yeasts consume sugar. They find this sugar all over the Earth, but mostly on plants and animals. In fact, yeasts are so abundant they can be found everywhere. There are probably yeasts floating around your house right this very minute!

When yeasts have a good food source, they grow and multiply very quickly. First they form a bud, which is really just a new cell that pops up on the parent cell. These buds grow into new yeasts and form long chains of yeast that are attached to the original yeast cell. These chains can form in less than two hours. Yeast is one fast-growing fungus!

Yeast breaks down the cells of plants and animal tissues, feeding on the sugar and other compounds produced while it does this. So, yeasts work as natural decomposers in our environment.

Although there are many different kinds of yeasts, let's study a few that are common in our lives.

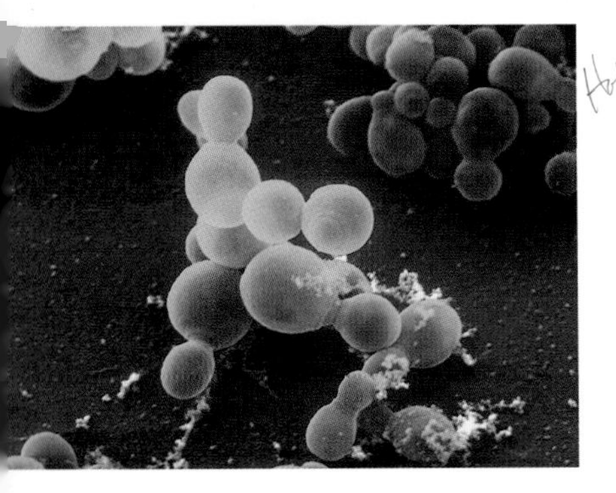

PARASITIC YEAST

One kind of yeast, known as *Malassezia*, lives on our skin. Occasionally it can become a parasite. A **parasite** is any living organism that lives off another living organism without giving anything in return. This parasitic yeast is usually a perfectly safe type of yeast. However, sometimes it changes into another kind of more harmful yeast that causes skin problems. When a parasitic yeast is harmful to the organism it lives on, we call it a pathogen.

So what kind of pathogenic problems can this yeast cause? *Malassezia* can cause dandruff, which gives people a flaky scalp. It can also cause dry skin or eczema to become worse, causing the skin to lose its pigmentation (meaning the skin turns white). Another kind of *Malassezia* can cause patches of darker skin. In addition, it can cause a condition known as ringworm, which isn't a worm at all, but a ring-shaped rash. Thankfully, *Malassezia* can be cured with the right type of antifungal medicines.

Another common parasitic yeast called candida also lives on humans. Candida is not a problem unless it reproduces too much. In that case, it becomes a pathogen and causes the person to get what we refer to as a yeast infection. This infection can form white patches in the mouth, called thrush. It can also cause itchiness in tender areas of the skin. Again, yeast infections are curable with medication. So even though parasitic yeast may hover around us, we need not be afraid of them.

BENEFICIAL YEAST

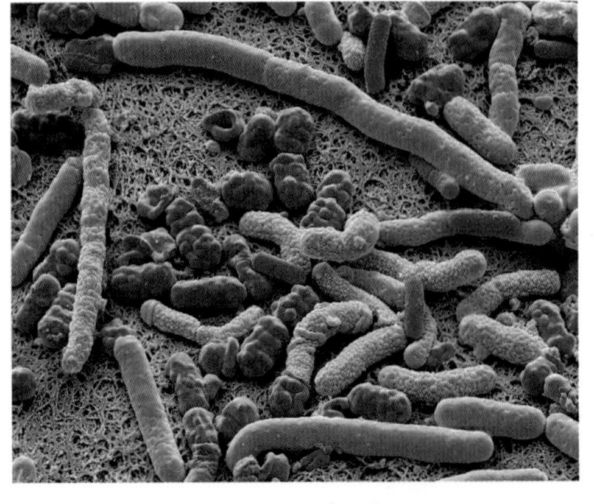

Of course, not all yeast is parasitic or pathogenic. Some are quite good for you. Have you ever heard of probiotics? The word comes from the Latin roots, *pro* meaning for and *bio* referring to life. So, probioitics are prolife! Many **probiotics** are tiny organisms that help humans and animals have a healthier body. Typically, when we consider probiotics, we think of the healthy organisms living in our intestines that help us digest food better. Most probiotics are good bacteria. However, one strain of yeast, *S. boulardii*, is also a probiotic (although it might be considered a parasite since it lives inside the body).

So you see, yeast isn't all bad. In fact, the most popular kind of yeast is not a parasite at all. Have you ever baked bread from scratch? If you did, you probably added a special fungus called baker's yeast, or scientifically: *S. cerevisiae*, which means sugar-eating fungus. Baker's yeast is a good kind of fungus. When combined with warmth, moisture, and any type of sugar, the dormant yeast jumps into action and consumes the sugar.

Did you know that when a living organism consumes food, it always produces a waste product? When you eat, you produce waste as well. That's what bathrooms are for. Interestingly, when baker's yeast consumes sugar, its waste product is a gas—carbon dioxide, to be exact When you are making bread and you have added the flour that carbon dioxide gas gets trapped inside the dough, making little air pockets in the bread. This causes the bread to expand and become quite fluffy. As

It's actually a fungus that makes the bread we love eating light and fluffy.

you can see, yeast is what makes bread rise!

In addition to producing carbon dioxide, yeast also releases another waste product: alcohol. In fact, alcoholic beverages, such as wine and beer, are made by adding a yeast called a certain kind of yeast to grapes (for wine) or watered-down barley (for beer). The yeast turns these items into alcohol. Humans have been using yeast to make alcohol for many thousands of years.

So now that you understand how yeast works, what do you think would happen to your bread if you forgot to add a sweetener, like sugar or honey, to your dough? What would happen if you added extra sugar? Let's do an experiment to test these ideas.

Before you begin your experiment, tell someone what you have learned about yeast.

ACTIVITY 14.1

EXPERIMENT WITH SUGAR AND YEAST

You will need:

- 5 bottles
- 12 tsp yeast
- 10 tsp sugar
- 5 balloons
- Funnel
- Measuring cups
- Measuring spoons
- Warm water

You will do:

1. Using the funnel, add 2¼ tsp of yeast to each bottle.
2. In the first bottle, add no sugar.
3. Add 1 tsp sugar to the second bottle, 2 tsp to the third bottle, 3 tsp to the fourth bottle, and 4 tsp to the fifth.
4. Using the funnel, add one cup of warm water to each bottle.
5. Cover each bottle with your thumb and shake gently.
6. Stretch a balloon over the mouth on top of each bottle to cover the opening.
7. After two hours, notice which balloons grew more.

What do you conclude? What happens if you do not add sugar to your bottle? Does more sugar make the balloon grow bigger? I hope you enjoyed your yeast experiment. It's time to move on to molds!

MOLDS

Because there are more than 100,000 species of mold, it's even more abundant than yeast. Mold is in the air. It's on the ground. It's everywhere, and all around! Mold is a fungus made up of many different kinds of cells. When mold cells find a suitable substance to eat and are provided the right amount of moisture and warmth, they'll reproduce and grow very quickly.

MENACING MOLD

Mold can be a bit of a menace. At first glance, you may not realize something has mold on it. But once it starts reproducing, it creates a furry layer. Have you ever pulled out a slice of bread from the bag and found little greenish spots on it? Perhaps you've left something in the refrigerator too long, and when you took it out you noticed a furry growth on the food. That's menacing mold! If we don't eat our fruits or veggies quickly enough, mold will start growing on them as well.

The furry layer growing on this strawberry is a type of mold.

Why do we find this menacing mold growing on our food? It's necause mold is a consumer and it consumes plant and animal matter. As it consumes, it also decomposes. So, the mold breaks down the food and then consumes it. Because of this, mold is an important decomposer. You see, mold is actually beneficial for the environment as a decomposer. But, if mold begins growing indoors where we live, it becomes a little problem or, in some cases, a big problem.

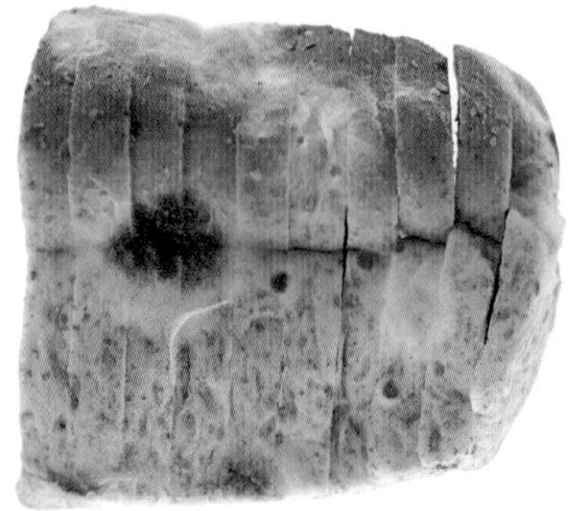

This loaf of bread is unsafe to eat because it's being consumed by various molds.

Although there are a lot of different kinds of molds, the most common molds found in a house are called *Aspergillus*, *Cladosporium*, and *Penicillium*. When you see a beam of light coming through the window, if you get in the right position, you may notice lots of dust in that stream of light. In some of that dust are the spores of the molds I just mentioned. Let's look at the first two of these molds for a moment. Later, we'll explore what we've discovered about penicillium.

An up-close look at mold growing on this tomato reveals a group of tiny individual structures.

Aspergillus grows on food and in air conditioning vents. When it collects in vents, it gets pumped in the air throughout the whole house. This is probably why food left in your room grows the same kind of mold as food left in the kitchen. This kind of mold can produce allergies as people breathe in the mold spores floating through the air.

When the common mold called cladosporium reproduces and begins to grow, it's often black or green and can be found growing inside our toilets and also on moist cabinets and walls. This mold isn't harmful to our health, but it sure is yucky.

There is one kind of very harmful mold that can grow in people's houses, causing many health problems. Thankfully it's not very common. This mold is called *Stachybotrys*. People often call this type of fungus black mold. But black mold isn't a good name for this dangerous mold because many molds that are not harmful at all are black. Therefore, just because you have a mold growing in your home that is black in color doesn't mean it is stachybotrys. Most likely it's not.

Stachybotrys grows in places where a great deal of moisture is left to sit. When *Stachybotrys* is

This worker is properly removing mold to prevent potential health problems to the homeowner.

found in a home, special companies with trained workers are called in to remove it. If it's not removed correctly, the spores can be spread to other parts of the home. No one wants a case of *Stachybotrys*. Because it grows in very moist environments, the best way to avoid ever having it is to make sure all water leaks, such as leaky pipes, are fixed quickly.

Hart As you can see, molds can be a problem for humans. In fact, one kind of mold actually changed the entire course of the history of the world. Yes, indeed. A mold caused the Great Famine in Ireland. This mold, sometimes called potato blight, completely wiped out all the potato crops in Ireland in 1845. Because the Irish were so dependent on potatoes, millions of people died. Of those that didn't die, most left Ireland and migrated to the United States. During that time, the population of Ireland went from eight million people to four million people. Can you imagine a simple mold having such an effect on an entire country and even the world?

Harleigh Molds have been the cause of other problems as well. Not long after the Great Famine, another kind of mold destroyed almost every grapevine in France. Before 1870, France was the leading producer of wine. But after the mold disaster, American vineyards began producing the most wine. Researchers aren't certain why, but American vines were hardy and able to resist the growth of this mold. As a result, California became famous for its wine production. This mold changed the world, at least as far as wine is concerned.

Harlan As you can see, molds have been quite a problem for humans throughout history. But we aren't the only ones to have suffered mold's effects. Mold is the reason for the extinction of a certain kind of amphibian called the golden toad. In 1989, scientists discovered that these strange lesions were forming on the skin of these toads. Within two weeks of contracting the lesions, the frogs would die. It turned it out these lesions were a kind of mold. Before long, all the golden toads were gone from the world. That's pretty sad, isn't it?

The golden toad became extinct due to a certain kind of mold.

MIRACULOUS MOLD

Dad Although mold is not usually our friend, in some cases, mold can save our life. You see, thousands of years ago, the ancient Egyptians discovered that if a moldy piece of bread was applied to a wound, it prevented the wound from getting infected. The reason wasn't well understood until many years later.

Here's how it happened: A scientist by the name of Alexander Fleming was studying bacteria. Do you remember that penicillium is one of the molds commonly found floating around our homes? Well, when Fleming came back from a trip, he discovered penicillium growing on one of his petri dishes containing bacteria. After examining it under a microscope, he realized that the bacteria in that dish had quit growing. It

Mycologists study mold by growing it in petri dishes like these.

turns out that some molds can kill bacteria. And that was the discovery of the antibiotic drug called penicillin that cures bacterial infections. Before then, bacterial infections almost always killed a person. Now, they are rarely fatal. Since Fleming's discovery, many kinds of molds have been made into antibiotics to cure many different kinds of infections.

Even though mold is typically not something you should consume, as you can see, there are some exceptions to this. Besides consuming antibiotic molds such as penicillin, we can safely consume molds found on a few kinds of cheeses. These special cheeses are made by adding mold to them during the production of the cheese. Blue cheese is one kind of cheese that is made with mold that's safe to eat. It's important to remember that not all cheese is made from mold, and that just because mold is growing on the cheese, it doesn't mean it's edible.

If you like the flavor of Roquefort cheese, you can thank a mold!

Tell someone what you have learned so far about molds before doing the mold activity below.

ACTIVITY 14.2

TEST MOLD ENVIRONMENTS

Let's learn more about the molds that grow on food. You know that molds need moisture to reproduce. Will adding extra moisture to a slice of bread cause mold to grow faster? What if the bread were actually quite wet? Would that enable the mold spores to reproduce even more rapidly? Let's find out.

You will need:

- 3 slices of bread
- Water
- 3 plastic Ziploc® bags
- Permanent marker
- Ruler

You will do:

1. With the marker, label the plastic bags: Dry, Moist, and Wet.
2. Place the first slice of bread into the bag labeled Dry.
3. Drop of few splashes of water on the second slice of bread. Place it in the bag labeled Moist.
4. Dip the third slice of bread in a bowl of water. Don't let it sit too long or it will fall apart.
5. Place the wet bread into the bag labeled Wet.
6. Seal and tape all three plastic bags closed. Do not open the bags, as you do not want those mold spores spreading around the house or entering your nose.
7. Place the plastic bags near a warm, sunny window.

8. With the ruler, measure the mold growth each day. (Measure from the outside of the bag. Do not open the bag.) Record the results on your Mold Growth Chart in your Botany Notebooking Journal.
9. Throw away the sealed bags when the experiment is finished.

Were you surprised by what you learned about mold growth in different environments? Write in your Botany Notebooking Journal your findings.

MUSHROOMS

You've learned a lot so far about fungi. Now it's time to explore the fungi called mushrooms!

For thousands of years the Chinese have been using these reishi mushrooms for medicinal purposes.

For thousands of years, mushrooms have been an important part of society. They have many uses in different cultures and have been beneficial around the world throughout time. In fact, a man was once found frozen in the Alps on the border of Austria and Italy. Scientists believed he lived more than 3,000 years ago. Can you imagine that? His frozen body was discovered 3,000 years after he died. When scientists searched this man's belongings, they found he was carrying with him two different kinds of mushrooms. One of those kinds of mushrooms is used to ignite fires. The other is now known to have antibacterial healing qualities. Did this man know that? We aren't sure. But we do know that the Chinese have used mushrooms for thousands of years to heal all kinds of ailments.

The mushrooms for sale at this roadside stall in Lithuania are filled with many essential nutrients.

But mostly, people enjoy mushrooms because they taste so good. A friend of mine lived in Russia as a missionary. Once, she was invited to dine at the house of a Russian friend. When she sat down to eat, she realized the entire meal was made up of many different kinds of mushrooms prepared a variety of ways. That's right! A mushroom meal! I'm sure it was quite delicious, and nutritious, too! You see, mushrooms are loaded with vitamins and minerals, as well as proteins and carbohydrates. And it's no wonder since they feed off plant and animal matter.

There are thousands of different kinds of edible mushrooms, and they grow all over the world. Because of this, mushroom hunting is a popular activity. Books have been written about mushroom hunting with tips on where to look for mushrooms and how to tell the good ones from the bad. People everywhere search the forests and fields for

These edible mushrooms are the trophies at the end of a forest mushroom hunt in Ukraine.

mushrooms, seeking that perfect mushroom meal. Many states have mushroom organizations with special trips to go mushroom hunting together.

Now it's vital that you understand not all mushrooms are edible. In fact, some are quite poisonous and can even be deadly. Some poisonous mushrooms look just like edible mushrooms, so it's important not to assume you've found an edible mushroom unless you know for sure. Though most poisonous mushrooms will only make you sick if you eat them, it's better to be safe than sorry. If you go mushroom hunting, be sure to bring a guidebook to help you identify the edible from the poisonous mushrooms.

Make sure you can identify which mushrooms are safe to eat before going on a mushroom hunt.

As you know, mushrooms aren't just good to eat. They have also been used in some unusual ways throughout history. During World War II, soldiers attached special glowing mushrooms to their helmets so they could keep track of each other in the dark. Glowing mushrooms have a special quality called bioluminescence. That means, like a firefly, they glow with a natural internal light.

Bioluminescent mushrooms are a magnificent creation of God!

Scientists believe some mushrooms glow in order to attract insects who will spread their spores.

There are too many different kinds of mushrooms and different uses for them to explain in this book. However, all mushrooms have certain things in common. Let's find out what they are.

MUSHROOM LIFECYCLE

When you hear the word mushroom, you probably think of an umbrella-shaped structure popping up from the ground. To be sure, many mushrooms are shaped like that. However, there is great diversity in the mushroom world, and they come in many different shapes and forms.

The part of the mushroom that you see above the ground is actually not the entire mushroom; it's simply the fruit of the mushroom! The rest of the mushroom exists below the surface. We call the fruit we see the fruiting body. Let's explore the anatomy of a mushroom.

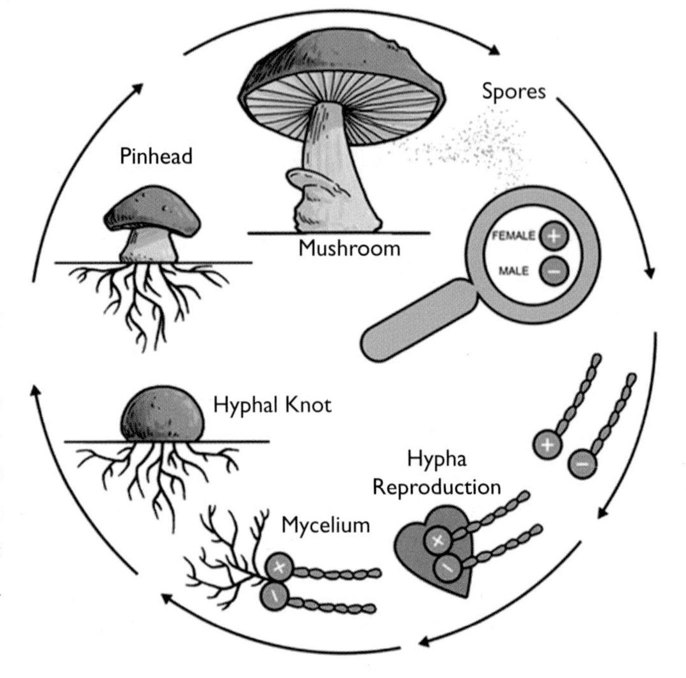

HYPHA

Just like the fungi we studied earlier, mushrooms produce spores. When a mushroom spore lands in a new location, if the conditions are just right, the spore begins to germinate. It produces tiny threads called **hyphae** that grow outward from the spore seeking to mate. These threads are usually so thin they're individually invisible to the human eye. The outer wall of the thread is made of **chitin**, the same substance that makes up the outer body of an insect, called the exoskeleton.

As the hypha grows to find food and a mate, it stretches longer and longer. When it encounters food, it releases a digestive enzyme that breaks down the plant or animal matter so it can pass through the chitin wall and be consumed by the fungus. Essentially, the mushroom digests the food outside of itself before bringing the food inside. What the hypha prefers to ingest will depend on the different species of mushrooms.

The white structures in this image are the mycelium branching out in search of nutrition.

MYCELIUM

When the hypha encounters a suitable mate, the male and female hyphae fuse and begin to form masses of hyphae we call **mycelium**. The mycelium branches out seeking nutrition.

The mycelium might grow for a few weeks, or it may grow for years. It can span a few feet or several hundred feet. Believe it or not, researchers discovered a mass of honey mushroom mycelium that spans more than 2,000 acres in the Malheur National Forest in Oregon. It's believed to be around 2,000 years old. That's a big ol' mushroom!

When the mycelium runs out of food, or there is a dramatic change in the environment—such as warmer temperatures or more moisture—it will form a hyphal knot that will then produce the fruiting bodies.

The fruiting bodies are like the fruit on a tree. If you see a cluster of mushrooms close together, they are probably all related to the same mycelium.

SPOROCARP

Typically, when a mushroom fruits, the first thing to pop up is a little pinhead that expands into a little button called a *primordium*. The primordium then pushes up from the ground forming a little stalk with a cap called a *sporocarp, or fruiting body*. In fact, the mushroom is often called a **sporocarp**. The Greek word *carp* refers to *karpos*, which means harvest. Considering all you've learned, what do you think the purpose of the mushroom is? Yes indeed. A mushroom's job is to make more mushrooms. As you probably figured out, the fruiting body contains the spores that will be released into the environment when the mushroom matures These fruiting bodies come in all different shapes and sizes.

Let's now explore some of the most common types of mushrooms.

This mushroom's spores are housed underneath the cap inside the gills.

GILL MUSHROOMS

The gill mushroom is the most easily recognizable mushroom. It has a stalk and a little umbrella-like cap with gills underneath the cap. The gills are called lamellae, and millions and millions of spores line the surface of the lamellae.

This indigo milk cap is beautiful to behold and delicious to eat when grilled or added to a stew.

BOLETES

Boletes look similar to gill mushrooms with a stalk and cap, but they do not have gills. Instead, they're equipped with thousands of little holes, or pores, underneath the cap. These pores are filled with spores.

Polypores look like stacked pancakes.

POLYPORES

If you see something like rows of pancakes growing on the sides of trees, you just might have found polypore mushrooms. These tough little structures are sometimes called bracket fungi because they look like little shelves. Like boletes, they have pores underneath that are filled with spores.

The stinkhorn is an interesting little mushroom that can benefit garden plants if the gardener is willing to put up with the stinkhorn's rotting meat smell.

STINKHORNS

These stinky mushrooms grow on a stalk with a little red head covered with spores. People often smell stinkhorns before they see them because of their strongly offensive odor. They give off this odor to attract flies who come for a visit then carry the stinkhorn's spores away to spread them far and wide. Plants can benefit from the stinkhorn because it breaks down rotting materials into nutrients the plants need.

These ball fungi are taking over the trunk of a tree.

BALL FUNGI

Ball fungi may or may not have a stalk. They are round like a ball, and their spores are usually found inside the ball. Some ball fungi mature by spreading out into a star shape to release the spores, and some release their spores through a pore at the top of the ball. Others simply crack open or disintegrate, releasing the spores as they harden and die.

JELLY FUNGI

These interesting fungi form a film or skin over dead plants or tree trunks. The mushroom is often soft and gooey and doesn't look much like a mushroom. But it is!

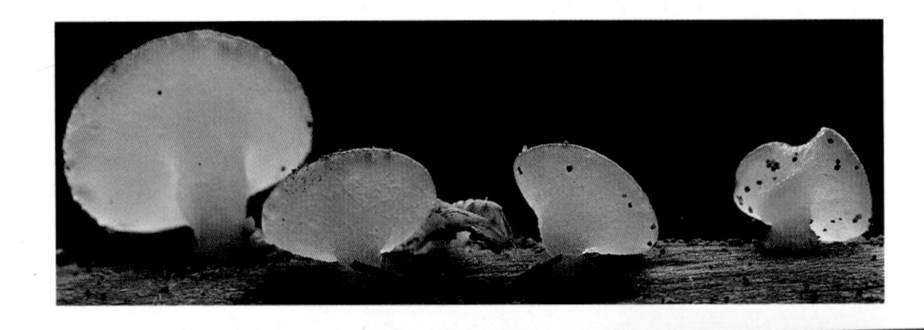

Jelly fungi come in all shapes and colors!

Do you see the little "eggs" inside the bird's nest fungi?

BIRD'S NEST FUNGI

Bird's nest fungi were named after their appearance, resembling the birds' nests we see in the spring. These little mushrooms often grow on trees or on mulch and look like little cups filled with eggs. Inside the "eggs" are the spores.

MORELS

Morels are edible mushrooms that many people enjoy. They have a cap that looks like an ocean sponge or even a honeycomb.

Morel.

It's easy to see how these cup fungi got their name.

CUP FUNGI

These fungi belong to the same group of fungi as yeast, so they are a bit different than the fungi mentioned above. Most of these fungi in the mushroom world form a little cup, and the spores are located inside the cup.

Before you move on, tell someone what you've learned so far about mushrooms.

DUST IN THE WIND

As you might imagine, it's important for fungi to get their spores some distance from the parent fungi. After all, they don't want to be competing for the same food.

The most common way fungi disperse their spores is through wind dispersal. But mushrooms are small and close to the ground; in order for a spore to get far away from the parent it must stay in the wind a long time. And in order to get blown about in the wind, the spore must stay dry. If the spore gets wet, it will become too heavy and fall right to the ground. Guess what? God designed fungi spores to stay dry! They actually resist absorbing water. We call them hydrophobic.

These puffball mushrooms are releasing their spores to be carried away to the perfect spot where they can become new mushrooms.

Most fungi spores, yeasts and molds included, are hydrophobic and can be blown a great distance from the original fungus. Because there millions of these spores are released, there's a great chance they'll find a place to land and begin growing.

The spores are like little dry bits of dust in the atmosphere, blowing around in search of a good landing spot. Once a spore finds the perfect location, it will let moisture in and germinate.

Because it's water resistant, you may be wondering what causes the spore to germinate. Scientists believe that when a spore lands on an appropriate food source, the spore recognizes the chemical makeup of the material. This chemical recognition causes the spore to germinate and develop the hyphae that will consume the food.

SPORE DISPERSAL

There are several different ways mushrooms release their spores. Most gill mushrooms and cup mushrooms use mechanical dispersal to eject their spores into the air. Here's how it works: Pressure is built up inside the mushroom as it grows. When the mushroom is fully mature, the pressure releases and the mushroom literally shoots the spores out into the air!

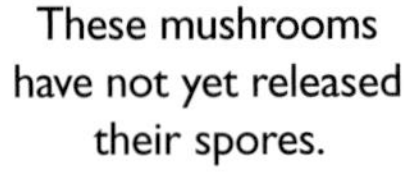

These mushrooms have not yet released their spores.

The mushrooms take on a new shape after releasing their spores.

Ball mushrooms often shoot their spores out through cracks or a hole in the top of the mushroom. When something impacts the mushroom, even something as small as a raindrop, the spores are pushed out of the ball mushroom. Bird's nest fungi also release spores when hit by a raindrop.

Though people find it extremely unpleasant, flies are attracted to the stinkhorn's strong scent.

Like I described earlier, the stinkhorn relies on its powerful odor for spore dispersal. The smell attracts flies and gnats that crawl around the stinkhorn, all the while getting covered in spores. They then fly off and take the spores with them to new locations.

Truffle mushroom spores are dispersed when animals eat them. These tasty treats are consumed by pigs and other animals. The spores are then released in the animals' droppings, spreading them from place to place.

Some spores may even be dispersed by water, carried in streams a great distance from the original fungus.

As you can see, like seed dispersal, there are many ways mushrooms disperse their spores. Once the spores are released in the wind, they can go a great distance. Scientists measured the distance traveled by spores of the fungus called wheat rust. The spores were carried more than a thousand miles in the wind. Fungi spores have also been found at great heights in our atmosphere. God designed these little water-resistant spores to go far and wide!

Tell someone all that you have learned so far about spore dispersal.

SAPROTROPHS

As you know, mushrooms, like all fungi, are consumers. Many mushrooms consume dead or decaying plant and animal matter. When an organism lives off dead matter, we call it a **saprotroph**. Imagine the amount of dead matter on Earth, with leaves and branches falling and plants and animals dying constantly. What if all that matter just stayed where it was for the rest of time. Earth would be quite a mess, wouldn't it? Thankfully, God thought of everything and designed many saprotrophic organisms to help clean up the Earth. Mushrooms are one of these saprotrophic

God showed His care for creation when He created saprotrophs to help clean up Earth.

organisms. They help Earth decompose by breaking down living things. This means mushrooms are beneficial for your garden!

FAIRY RINGS

One can easily see that mushrooms are good for plants when they spot a fairy ring. What's a fairy ring? I'll tell you. Sometimes the mycelium grows fruiting bodies in a little circle around the perimeter of the mycelium. We call these circles of mushrooms fairy rings. All the mushrooms in that circle are from the same mycelium. Fairy rings often occur on patches of grass. Scientists have noted that the grass inside the circle is healthier than the grass outside the fairy ring. Why is that? It's because under the surface of the grass, inside the circle, the mycelium is decomposing the plant matter, turning it into healthy compost, providing the grass with more nutrients.

So when you see mushrooms pop up in your garden, be glad. They're probably helping your plants!

PARASITIC MUSHROOMS

Although many varieties of mushrooms are beneficial to our gardens, some mushrooms are not. These harmful types of mushrooms are parasitic, living off the nutrients of a plant without giving anything in return—eventually killing the plant.

This parasitic mushroom is stealing important nutrients from the tree.

Parasitic mushrooms often grow on trees, taking the nutrients the trees need to survive. Usually, these mushrooms only attack trees that are already weak and sick. However, a weak tree that's being consumed by these mushrooms can spread the mushrooms to nearby healthy trees, causing them to grow weak and eventually die as well. So it's important to get rid of these mushrooms as soon as you notice them invading your trees.

However, once the parasitic mushroom has killed the tree, it actually consumes the dead tree. That means the mushroom changes from a parasitic mushroom to a saprotrophic mushroom since it's now decomposing the dead plant matter.

There is a really odd parasitic mushroom that actually invades living insects, mostly caterpillars. This mushroom is called the trooping cordyceps. It attaches to the head of the insect, sending its hypha into the insect where it grows a mycelium, eventually killing the creature. Once the creature is dead, fruiting bodies develop.

You can tell by the little yellow fruiting bodies that a trooping cordyceps has invaded this moth.

Interestingly, some of these parasitic mushrooms, such as the trooping cordyceps mentioned above, can actually be beneficial. In some instances they are used as medicines or food for humans.

As you've learned, some mushrooms make the soil healthier, but some aren't just good for the soil, they're fundamental to the plant's health! Let's take a look at these important mushrooms.

MYCORRHIZAL MUSHROOMS

Mycorrhizal mushrooms are some of the most important mushrooms in the world. They form a relationship with plants that help both to survive. It's a mutually beneficial relationship. Do you remember what we call a good relationship between two organisms? Symbiosis! Mycorrhizal mushrooms are symbiotic with many trees and plants.

Orchids have a symbiotic relationship with russula mushrooms like these.

Would you be surprised to learn that around 95% of plants benefit from this symbiotic attachment to fungi? Plants that form these relationships develop into larger, stronger, and faster-growing plants than the ones that don't. Some trees would not be able to survive at all without a fungus growing around their roots. In fact, all orchids rely on mushrooms to germinate. Do you remember that orchid seeds have no endosperm? Because of that, they rely on mushrooms to provide the seed with the nutrients it needs to germinate. Most of the mushrooms at help orchids survive form jelly-like substances. However one kind, the Russula, gill mushroom fruiting body. Without these important mushrooms, there would ids! As you can see, mychorrhizal mushrooms are an essential part of nature.

So how do these mycorrhizal mushrooms help the plant? You may remember that *myco*, refers to mushrooms, and that *rhiz* refers to roots. This is a good name for these mushrooms because they join forces with the roots of certain plants by weaving their mycelium into the root cells and wrapping all around the roots of the plant. Why is this beneficial? Because once the roots are wrapped, the mycelia can then provide moisture, phosphorus, and other important nutrients to the plant.

So what does the mushroom get out of this arrangement? It feeds off the sugars that the plant produces. Everyone's healthy and happy!

Wow! It's hard to believe you've completed your studies of botany and mycology. It's been a long and fruitful journey for sure! Just think of all the things you learned that you didn't know when you started this course. So much of what you now know will come in handy throughout your entire life. Just remember, like the organisms you studied, God wants you to be fruitful on this Earth. I know you will be if you continue to keep your faith strong by filling your heart with all that is good for you. Just like plants need nourishment to grow strong and healthy, so your faith needs tending to as well. You can do this by spending time in prayer, reading the Bible, and worshiping God.

Let's complete our studies by finishing our Botany Notebooking Journal and then heading out on a mushroom hunt!

WHAT DO YOU REMEMBER?

What are the three kinds of fungi? Name a kind of yeast that is harmful. What products do yeast release to make bread rise and to make wine or beer? Which kind of mold was first discovered to cure bacterial infections? Where is mycelium found? What produces fruiting bodies? What is the purpose of the fruiting body? Why is it important that spores are hydrophobic (water resistant)? Name a few ways mushrooms release or disperse spores. Why are most mushrooms good for your garden? What are mycorrhizal mushrooms? How do these mushrooms benefit plants? What does the mushroom receive from the plant?

ACTIVITY 14.3

MYCORRHIZAL MISSION STORY

I'd like you to write a story about a Russula mushroom spore who is on a journey in search of an orchid seed. Its mission is to find this special seed so that a new orchid plant can grow. What obstacles will the spore encounter on its journey? What happens when it finds the seed? You can make your story with only pictures or only words, or you can do both. Write your story in your Botany Notebooking Journal.

ACTIVITY 14.4

HUNT MUSHROOMS

Take your nature journal with you and go on a hunt for mushrooms! Look in forested areas and near streams, rivers, and any place that is moist. Look on trees and on the ground. When you find a mushroom, make a sketch of it in your journal. When you get home, do some research to find out the species of mushroom you saw. Wash your hands after you handle them. Remember not to eat any mushroom.

ACTIVITY 14.5
GROW EDIBLE MUSHROOMS

Many types of mushrooms are nutritious and tasty and can be used in a variety of recipes. Let's grow some that are safe to eat!

You will need:

- Store-bought mushrooms with stems in place
- Brightly colored cardstock (don't use black or white as some spores are black or white)
- Plastic cup
- Pitcher of water
- Woodchip mulch (do not use pine mulch or cedar mulch)
- Plastic tub with lid
- Colander
- Compost
- Scissors
- Spray bottle filled with distilled water

You will do:

1. Prepare a substrate in which to grow your mushrooms. First, put two inches of compost in the bottom of the plastic tub.
2. Put mulch in the colander and place it in your sink.
3. Heat the water in a microwave or on the stove to boiling.
4. Pour the water over the mulch to sterilize it.
5. Pour the sterilized mulch into the tub of compost.
6. Mix the mulch and compost thoroughly. Put the lid on the tub.
7. Now you need to get some mushroom spores. Knock out the stalk in the center of the mushroom and carefully remove it from the mushroom.
8. Place the mushroom gill side down on the colored paper.
9. Place the cup over the mushroom.
10. In 48 hours, take off the glass and remove the mushroom to reveal the spore prints released from the mushroom.
11. Cut up the spore prints with scissors to make several small squares that have spore prints on them.
12. Spread the spore print pieces throughout the tub of compost-mulch substrate and mix thoroughly to cover the spore print pieces with substrate.
13. Put the lid back on the tub.

14. Mist the substrate daily with distilled water to keep the humidity in the box high.
15. Make sure the mushroom tub is stored in a place where the room temperature is 60–75 degrees.

In a few weeks, you should have mycelium growing. Watch for new mushrooms popping up soon after and add them to your favorite recipes!

APPENDIX

SUPPLY LIST

Lesson 1

- Cover paper (construction paper, colored card stock, or scrapbook paper)
- Copy paper (10–12 sheets)
- Stapler
- Stack of cardboard (or substitute a kitchen cutting board)
- Standard-sized paper towels
- Red, blue, and yellow food coloring
- 5 small, clear cups
- Spoon
- Water
- Large, empty, open cardboard box
- Aluminum foil
- Single socket pendant lamp cord (for lanterns)
- LED full spectrum grow light bulb
- Glue
- Scissors
- Seeds
- Small flowerpots
- Vermiculite
- Peat moss
- Compost
- Your light hut
- Water
- Timer (optional)

Lesson 2

- Seeds (bean and sunflower work best)
- 3 plastic Ziploc® bags
- 3 paper towels
- 3 or more turnip seeds (or bean seeds)
- Tape
- Ruler (should read centimeters)
- Scientific Speculation Sheet

Lesson 3

- Flower (A lily works best because all parts of the flower are easily visible.)
- Flower dissection page of your Botany Notebooking Journal
- Glue or tape
- Adult to use a knife in this activity
- Cup of soil
- Sunny windowsill
- Sunflower seed
- Fresh flower
- Container with a lid
- Borax

Lesson 4

- 2 flowers that are still on their plants. They should be on separate plants of the same type.
- Cotton swab (like a Q-tip®)
- Small mason jar with lid
- Red spray paint
- Adult with a drill or hammer and nail (to punch holes in the lid)
- 1 yard of wire (cut in three equal pieces)
- 1 small piece of wire (for fastening other wires)
- Water and sugar
- Vermiculite
- Compost
- Peat moss
- Flowering plants that produce the kind of nectar most butterflies enjoy eating (butterfly bush, lantana, zinnia, bee balm, purple coneflower, penta, sage, milkweed or butterfly weed, lilac, sunflower, marjoram)

Lesson 5

- 1 Really ripe banana
- 1 fresh banana
- Squash or medium-sized pumpkin
- Adult with a sharp knife
- Burr
- Magnifying Glass
- Apple cut into 5 slices
- 4–5 cups filled with different substances: lime juice, vinegar, olive oil, water, and/or saltwater. You can also add more cups and try other substances you think may be helpful in preserving the fruit's color.

Lesson 6

- 2 identical candles
- Matches or lighter
- Adult
- 2 large jars
- Small plant
- Plank of wood
- Large glass jar
- Organic potato (regular potatoes are sprayed with a substance to keep them from growing roots)
- 2 bamboo skewers
- Water
- Plastic sandwich bag
- Clothespin
- Living plant that is not an evergreen
- Mod Podge®
- Card stock or blank canvas
- Fall leaves
- Paintbrush
- Multiple different leaves

Lesson 7

- Black bucket with lid
- Adult
- Hammer and nail (or drill)
- Compostable material
- 5–8 paper towels
- Water
- 5–10 bean seeds
- Romaine lettuce scrap (the bottom two inches of the base)
- Celery scrap (the bottom two inches of the base)
- Glasses or jars
- Planter big enough for two vegetables
- Garden soil

Lesson 8

- Celery
- Cup
- Water
- Blue food coloring
- Clay or Play-Doh®
- Yellow, green, red, and blue food coloring
- 4 tall glasses
- 4 white roses or white carnations (If neither of these flowers are available at your local store, you can try to use another variety of white flower.)
- 2 paper or Styrofoam® cups with a lid
- 4 bean seeds
- Mixture of vermiculite and compost for soil
- Sharpened pencil
- Black paint

Lesson 9

- Materials to build a raised garden bed
- Compass
- Empty plastic bottle with lid (sturdier energy drink bottles with a wide mouth work best)
- Drill or hammer and nail

Lesson 10

- Ruler
- Colored pencils
- Twig on a tree
- Someone to help you
- Measuring tape or yardstick to measure the height of your helper
- 12-inch ruler
- Tall tree
- Crayon with all the paper removed
- Plain white paper
- Tacks (optional)

Lesson 11

- Tape measure
- Chalk
- 2 plastic sandwich bags
- 2 clothespins
- Deciduous tree
- Conifer tree
- Pinecone
- Oven preheated to 250 degrees
- Bucket of cold water

Lesson 12

- Paper
- Several fern fronds
- Several colors of paint
- Paintbrush or sponge
- Large glass container with a lid
- Pea-Ver-Comp mixture
- Miniature fern

Lesson 13

- Coat hanger
- Yarn
- Tape
- Trees with lichens on them, preferably in different places (Lichens prefer oak trees because they can keep the water locked into the deep ridges of the oak bark better than in the bark of most trees.)
- 3 handfuls of moss
- 2 tablespoons of water retention gel
- Half cup buttermilk
- Blender
- Bucket
- Paintbrush (2 inch)
- Spray bottle with water
- Piece of plywood or another surface where you want your moss to grow (Ideally you want the surface to be located in a shady area.)

Lesson 14

- 5 bottles
- 12 tsp yeast
- 10 tsp sugar
- 5 balloons
- Funnel
- Measuring cups
- Measuring spoons
- Warm water
- 3 slices of bread
- Water
- 3 plastic Ziploc® bags
- Permanent marker
- Ruler
- Store-bought mushrooms with stems in place
- Brightly colored cardstock (don't use black or white as some spores are black or white)
- Plastic cup
- Pitcher of water
- Woodchip mulch (do not use pine mulch or cedar mulch)
- Plastic tub with lid
- Colander
- Compost
- Scissors
- Spray bottle filled with distilled water

WHAT DO YOU REMEMBER? **ANSWER KEY**

Your child should not be expected to know the answer to every question. These questions are designed to help you review important concepts with your child. The questions are in plain type; the answers are in bold , but please keep in mind that answers could vary and still be correct. We recommend you spend the most time talking about the final question in each lesson.

Lesson 1

Why do scientists use Latin? **Because Latin words and their meanings will never change.**

What is a biologist? **A scientist who studies living things.**

What is a botanist? **A biologist who studies plants.**

What are some helpful or interesting things that botanists do? **Study plants that are used to make medicines and cure diseases. Experiment with plants to help crops grow faster and stronger. Help farmers improve the food we eat. Study plants in outer space to see how gravity affects plants.**

What do vascular plants have that nonvascular plants do not have? **Tubes inside the plant that transport fluids, like water.**

What are spermatophytes? **Seed plants, or plants that produce seeds.**

Can you name one? **A pine tree**

What are sporophytes? **Spore plants, or plants that make spores instead of seeds**

Can you name a plant that is a sporophyte? **A fern**

Lesson 2: Seeds

What is a seed? **A baby plant in a protective covering**

What does dormant mean? **Asleep, from the Latin word dormire which means to sleep**

What does a seed need to wake up and begin growing? **Warmth, water, and air**

What is the baby plant in a seed called? **An embryo**

What is the seed's testa? **The seed's coat**

What does the testa do? **Protects the embryo**

What is the hilum on a seed? **The place where the seed was attached to its mother**

Describe germination. **Students should refer to the illustration on p. to describe the process of germination.**

What is the top part of the embryo called? **The epicotyl**

What are the feather-like leaves on the embryo called? **The plumule**

What is the embryo's root called? **The radicle**

What is the nutrition within the seed called before it gets absorbed by the cotyledons? **The endosperm**

Explain how the testa comes off for germination. **When the seed gets wet, water gets inside the testa making it soft and soggy and causing it to fall off.**

What is a producer? **A living thing that makes, or produces, its own food**

What is a consumer? **A living thing that eats, or consumes, other living things for food**

Are plants producers or consumers? **Producers**

Are people producers or consumers? **Consumers**

Lesson 3: Angiosperms

What is so special about angiosperms? **They provide us with the things we need, like food and clothing, and often produce beautiful flowers.**

What is the purpose of a flower? **To make seeds**

What is the job of the sepal? **To cover and protect the developing flower bud**

What are all of the sepals together called? **The calyx**

Explain what the corolla is. **It is all of the petals together.**

What is the male part of a flower called? **The stamen**

What is the female part of the flower called? **The carpel**

Where are the flower's ovaries? **At the bottom of the style on the carpel**

What do ovules become after they are pollinated? **Seeds**

Tell something you learned about flowers in the daisy family. **Answers will vary**

Tell something you learned about flowers in the orchid family. **Answers will vary**

Tell something interesting your learned about carnivorous plants. **Answers will vary**

Lesson 4: Pollination

What are the things that attract insects to flowers? **Smell, color, patterns (nectar guides)**

What color is the hummingbird most attracted to? **Red**

Name some mammals that pollinate plants. **Bats, lemurs, and some rodents**

Explain wind pollination. **Pollen is blown off the tree and glides in the wind until it lands on another tree of the same kind nearby.**

Explain self-pollination. **Pollen from a flower's stamen goes to the carpel on the same flower.**

Can you also explain why a flower petal dries up and falls off after it has been pollinated? **Because it has finished its work of attracting pollinators and can then spend its energy manufacturing seeds. Also, other flowers will get an equal chance to be pollinated which means plenty of seeds will be produced.**

Lesson 5: Fruits

What is the main purpose of fruits? **To serve as a container that protects the seeds as they mature and to help disperse, or spread, the seeds.**

What is the difference between a fruit and a vegetable? **A fruit is anything on a plant that serves as a container for the seeds. A vegetable is any edible part of a plant that does not have seeds.**

Name a few different kinds of fruits about which you learned. **Answers will vary: fleshy fruits, dry fruits**

Describe what seed dispersal means. **It means spreading or scattering, and is the process of getting the seeds from the parent plant to a new location.**

Explain the different methods of seed dispersal. **Human dispersal- when farmers plant seeds in the ground; Water dispersal- when water currents take seeds to new places; Animal dispersal- when animals get seeds on their fur or consume them to be carried and dropped in new locations. Wind dispersal- when breezes carry seeds to new places. Mechanical dispersal- when plants fling their seeds out into the open to be taken away.**

Lesson 6: Leaves

Why are the leaves of a plant so important? **Because they make food for the plant.**

Can you explain to someone what the stomata do for a plant? **They take in air to help make the food the plant needs.**

Explain what would happen if a plant lost all its leaves. **It would die of starvation because it would have no way of getting the chemicals from the air it needs to make food.**

What does a plant take in from the air and what does it put back into the air? **It takes in carbon dioxide and releases oxygen.**

What kind of food does the plant make? **Sugar**

Explain photosynthesis in your own words. **It's the process of a plant using energy from light to help put together carbon dioxide and water in order to make sugar and oxygen. The plant uses the sugar for food then releases the oxygen into the air for us to breathe.**

What makes leaves green? **Chlorophyll**

What are the four things a plant needs to make food? **Water, carbon dioxide, light, and chlorophyll**

What happens when one ingredient is removed? **The plant is unable to perform photosynthesis and will eventually die.**

Lesson 7: Roots

What are the three main jobs of roots? **To absorb nutrients and water from the soil, to hold the plant in place like an anchor, and to prevent erosion.**

Explain why root hairs are important. **Root hairs do most of the work of absorbing water and nutrients for the plant; without them the plant would die.**

Where do roots add to their length? **Roots grow longer from the tip, adding cells to the end of each root.**

What is the root cap? **The strongest part of the tip of the root. It's made up of a group of tough cells whose job is to push through dirt in search of moisture and nutrients.**

What are the roots always looking for? **Water and nutrients**

What is geotropism? **A root's special sense that tells it to turn and grow down into the Earth**

What is another name for geotropism? **Gravitropism, meaning turning in the direction of gravity**

Tell me about rooting a plant. **You take a healthy stem, branch, or even leaf from a plant and put it in soil or water to grow a new plant of the same kind. It is considered a clone, or copy, of the original plant because it has the same DNA.**

Lesson 8: Stems

What two parts of the plant are in the vascular bundle? **The xylem and the phloem**

Explain what xylem and phloem do for the plant. **The xylem are the tubes that send water up from the roots to the rest of the plant, and the phloem are the tubes that send the sugary food down from the leaves to the rest of the plant.**

What is the difference between woody and herbaceous stems? **Woody stems are hard and woody and grow thicker with age by adding cells to their vascular cambium. They cannot perform photosynthesis. Herbaceous stems are soft and green and can perform photosynthesis and feed the plant. They grow longer by adding cells to the end of the stem.**

What is phototropism? **When plants use special chemicals to make their stems bend and twist to turn toward the light.**

What are the special chemicals called that enable a plant to grow toward the light? **Auxins**

Lesson 9: Gardening

What are some good reasons to grow your own food? **To protect yourself from harmful chemicals found in most grocery store produce. To save money on your grocery bill. So you can eat more nutritious and delicious ripe fruits and vegetables.**

What zone are you in? **Answers will vary**

What is your best growing season? **Answers will vary**

Which side of the garden should you plant tall growing plants? **On the north or west side**

Name some tips for watering your garden. **Use a Pea-Ver-Comp mix to prevent your plants from drowning. In the hot summer months check the soil daily to make sure it's very moist. Thoroughly soak the soil each time you water the plants, aiming the water toward the roots and away from the leaves. Water in the early morning or late evening.**

Why is pruning important? **It allows more sunlight and airflow and enables the plant to put all its energy into producing the flowers and fruit.**

What are some solutions for getting rid of pests in your garden? **Prepare the appropriate mixtures to spray on the leaves of your plants, cover your plants with garden fabric, clean your gardening tools and containers, always use fresh soil, allow plenty of space between your plants, check your plants often and carefully to treat problems immediately.**

Name a solution for getting rid of diseases on your plants. **Apple cider vinegar and water**

Lesson 10: Trees

Name some important uses for trees. **They provide shelter, shade, beauty, food, and healthy air for humans and animals. Trees also keep Earth's landscape stable. They can lower crime rates, increase property value, help save energy by cooling temperatures, protect from the wind, and even raise students' test scores.**

How can you tell, by looking at a tree's branch, how much a tree has grown? **There are marks along the twig of the branch that tell us exactly how tall the tree grew in each previous year.**

Explain the anatomy of a twig. **Be sure to include terminal buds, lenticels, nodes, internodes, and auxiliary buds in your explanation. Students should refer to the illustration on p. to explain twig anatomy.**

Explain why words carved in a tree remain in the same location many years later. **Because a tree grows taller from the top by adding to the tips of its branches. A tree's trunk will not continue to grow taller, only thicker.**

Describe the layers, names, and functions of a tree's trunk. **Students should refer to the illustration on p. to describe a tree's trunk.**

Lesson 11: Gymnosperms

What are the three kinds of conifer leaves? **Needle-like, scale-like, and awl-like**

What is the process by which conifers produce seeds? **Conifers have two cones, the male pollen cone and the female seed cone, which contains eggs. When the pollen cone releases its pollen, some of it reaches the eggs in the seed cones. Fertilization then happens and seeds begin to form inside the seed cone.**

Why are cycads considered gymnosperms? **Because their seeds are made in open containers, cones, rather than closed fruits**

Describe a cycad. **It has a palm-tree trunk topped by a whorl of palm-like leaves. Its cones are produced in the center of the leaf whorl.**

Describe the ginkgo. **Its leaves are broad and flat like those of an angiosperm. It also loses its leaves in the fall and regrows them in the spring, which means it's deciduous.**

Why are ginkgos considered gymnosperms? **Because they produce uncovered seeds**

Why do people prefer to plant the male ginkgo biloba? **Because the female ginko produces seeds with a terrible odor**

Name the four kinds of forests. **Tropical rainforests, boreal forests, temperate deciduous forests, and temperate coniferous forests**

How can forest fires be beneficial to a forest? **The decayed remains of burned trees fertilize the soil, the cleared landscape can become home to a larger variety of plant species than was there before, and a controlled burning can promote the health of a certain part of the forest.**

Lesson 12: Seedless Vascular Plants

What kind of roots do ferns have? **Rhizomes**

What is a frond? **The leafy blade of the fern, or a fern leaf**

What are the spots sometimes found on the underside of the frond? **Sori, which are clusters of sporangia**

In the fern lifecycle, what is the little structure with male and female plant parts? **The prothallus**

Can you name the male part? **Antheridia**

Can you name the female part? **Archegonia**

What must be present for the sperm to reach the egg? **Water**

What is the little baby frond that unfurls called? **A fiddlehead**

What are some other ways ferns produce new plants? **Ferns can root by spreading runners along the ground to grow new ferns at each new rhizome location; they can self-root when a frond touches its tip in the dirt and a new fern sprouts; the prothallus can make a gemma which is carried off to a new place to produce a new fern.**

What are tree ferns? **Giant ferns that look like palm trees and cycads**

How are they different from cycads and palms? **Their trunks are made of a mass of roots, they have fiddleheads, and they reproduce with spores.**

Lesson 13: Non Vascular Plants

How do nonvascular plants distribute moisture and nutrients? **By diffusion, a process where the plant absorbs water and nutrients that will soak through to other parts of the plant.**

What does it mean when a plant desiccates? **It dries out for long periods of time without actually dying.**

Where do the sporophytes grow once a moss plant is fertilized? **Inside the archegonium**

Describe the two types of liverworts. **Leafy liverworts look like leaves on a stem lying horizontally on the surface of the ground. Thallose liverworts look like a mass of flattened or wavy green tissue on the ground and have a cup-like structure that contains their sperm.**

What two organisms combine to make lichen? **Fungi and algae**

What are some uses for lichens? **They protect trees and provide clean air to breath, food for animals, homes for insects, nesting material for hummingbirds, dye for fabrics, and at one time were used for medicines and poison for the tips of Native American arrowheads.**

Lesson 14: Mycology

What are the three kinds of fungi? **Yeast, mold, mushrooms**

Name a kind of yeast that is harmful. **Malassezia, candida**

What products do yeast release to make bread rise and to make wine or beer? **Carbon dioxide and alcohol**

Which kind of mold was first discovered to cure bacterial infections? **Penicillium**

Where is mycelium found? **Underground**

What produces fruiting bodies? **The mycelium**

What is the purpose of the fruiting body? **To reproduce**

Why is it important that spores are hydrophobic (water resistant)? **So they can stay dry and be blown far away from the parent fungi**

Name a few ways mushrooms release or disperse spores. **Mechanical dispersal, smell, animal dispersal, water dispersal**

Why are most mushrooms good for your garden? **They decompose plant and animal matter, turning it into a healthy compost**

What are mycorrhizal mushrooms? **Mushrooms that form a symbiotic relationship with plants**

How do these mushrooms benefit plants? **They grow around the plant's roots, providing them with moisture, phosphorous, and other important nutrients.**

What does the mushroom receive from the plant? **Sugar**

INDEX

N

O

P

R

S

T

U

V

W

XYZ

PHOTO CREDITS

Unless otherwise indicated below, all photos and illustrations are either created by the author or designers, are in the public domain, or are licensed stock photos from GettyImages.com, Alamay.com, Pixabay.com, and Dreamstime.com.

Lesson 4
Honey Possom: ©Fredy Mercay/ANT Photo Library/Science Source, 63; Bat: © Merlin D. Tuttle/ Science Source, 70; Flying Fox: Andrew Mercer (CC BY-SA 4.0), 70.

Lesson 7
Root hair: ©Steve Gschmeissner/Science Source, 114; Geotropism: ©Martin Shields/Science Source, 117.

Lesson 13
Lichen: ©Alex Proimos (CC BY-A 2.0), 194;

Lesson 14
Yeast: ©Biophoto Associates / ScienceSource, 201; Practical Yeast: ©ScienceSource, 202; Beneficial Yeast: © Steve Gschmeissner/Science Source, 202; Mushroom Market: Phillip Capper (CC BY-A 2.0), 207; Morel: Morchella_conica_1 (CC BY-SA 1.0), 211;